Frontiers of Biomechanics

Volume 5

Biomechanics is a rapidly expanding field that focuses on exploring the mechanics of structure–function relationships underlying the functionality of living systems with the aim of gaining insights for medical applications. The field cuts across diverse disciplines: physics, biology, medicine, mechanical engineering, and biomedical engineering, and incorporates principles and concepts from these fields in the study of living systems and their dynamics.

A major objective of this series is to pioneer new frontiers of biomechanics by including concepts and ideas from new research frontiers yet to be addressed within the existing paradigm of biomechanics. The series highlights the role of mechanics in the realization of multicomponent and hierarchical interactions that characterize living systems, based on the context of open systems. The series also presents in-depth coverage of cutting-edge research in frontiers of biomechanics that explores fundamental concepts of what constitutes life, based on mechanics as an entity of energy principles and information science. It is hoped that this series will develop as an open system (analogous to living systems) that is admissible of future research dynamics and expansion in the field of biomechanics.

Distinctive features of the series:

- Each volume will cover core topics in the frontiers of biomechanics.

- The series will be multidisciplinary in that it will consist of volumes written by authors who are at the forefront of research in diverse fields including physics, mechanical engineering, biology, and medicine.

- Each volume will place emphasis on not only providing readers with new knowledge and techniques in biomechanics, but also presenting the underlying concepts and principles concisely and precisely so that the readers are able to discern unresolved and challenging problems in biomechanics.

– The series aims to contribute to further expansion and development of the field of biomechanics by including concepts and ideas from new research frontiers yet to be addressed within the existing paradigm of biomechanics.

Jiro Nagatomi

Mechanobiology of Hydrostatic Pressure

Frontiers of Biomechanics

Springer

Jiro Nagatomi
Department of Bioengineering
Clemson University
Clemson, SC, USA

ISSN 2199-8515 ISSN 2199-8523 (electronic)
Frontiers of Biomechanics
ISBN 978-981-95-4404-2 ISBN 978-981-95-4405-9 (eBook)
https://doi.org/10.1007/978-981-95-4405-9

This Springer imprint is published by the registered company Springer Nature Singapore Pte Ltd.
The registered company address is: 152 Beach Road, #21-01/04 Gateway East, Singapore 189721, Singapore

Preface

Biomechanics is the development, extension, and application of mechanics for the purpose of understanding better the influence of applied loads on the structure, properties, and function of living things and the structures with which they interact.
-J.D. Humphrey

Above is the definition of biomechanics provided in a popular textbook, *An Introduction to Biomechanics* (Humphrey and O'Rourke, Springer Science + Business Media, New York), which I use to teach my undergraduate course. After presenting this long but comprehensive definition by a world-renowned biomechanics professor, I tell students that biomechanics can mean different things to different people. Then, to illustrate the point, I show students the program from the World Congress of Biomechanics, a five-day conference that takes place every four years on a different continent/region. The pages of the scientific program list the oral and poster presentation sessions categorized into a wide range of biomechanics subtopics including, but not limited to, human movement, blood rheology, brain injury, etc., indicating that researchers apply many different areas or mechanics to many different organs, but they are all called biomechanics. Out of approximately 300 sessions at the most recent conference in 2022, twenty of them included the term "Mechanobiology" in the session topic, indicating significant interest among the biomechanics community. Just like biomechanics, the definition of mechanobiology may also vary depending on whom you ask. However, I consider mechanobiology "studies of the effects of mechanical environments on the biological processes of cells." What I mean by "mechanical environments" are the forces that result from any physiological function of the human body, like pumping of the heart or loading on the skeleton through daily activities, which generate hydrostatic pressure, fluid flow, and tension on and around the cells of any organs. Although most cells experience not just one type of mechanical loading but combinations like hydrostatic pressure and stretch, hydrostatic pressure and flow, or all three, this volume will focus on the mechanobiology of hydrostatic pressure.

I have been studying the effect of hydrostatic pressure on cells for almost three decades. The reason why I got into it in the first place as a graduate student researcher was quite simple. Another student in my research group was already studying the effect of fluid flow on bone cells, so my advisor needed someone to start on hydrostatic pressure on bone cells. Looking back, I guess I was very naïve and didn't know how complex and controversial the mechanobiology of hydrostatic pressure was. However, I was excited to build a new experimental setup to apply hydrostatic pressure on cultured cells and to explore the unknown territory. When I first presented my data on the effect of cyclic hydrostatic pressure on osteoblast proliferation at a conference, I became aware that some researchers were entirely against the idea that cells can respond to small changes in hydrostatic pressure. In front of my poster, I was grilled by three professors from other universities as they looked for any reason to discredit my work. Similar encounters continued for many years, even after I became a professor. When my former student gave his research seminar on this topic for a job interview, someone in the audience got up and said, "Nice presentation, but I don't believe any of your results." Some people would not accept that hydrostatic pressure can impact cell behaviors. On a separate occasion, another world-renowned biomechanics researcher said to me politely, "I find it very difficult to understand how cells can sense small changes in hydrostatic pressure." Then I immediately replied, "Well, that is why we need to study it to figure it out." While there were some non-believers, there have been numerous reports on various types of cells responding to small changes in hydrostatic pressure. Those who published their work on the mechanobiology of hydrostatic pressure may have faced criticisms from the naysayers as much as I have. My graduate school mentor told me on more than one occasion, "People used to believe that the Earth was flat." That simple statement reminded me that in the history of science, there have always been controversies, opposing theories, and discrepancies in experimental findings. What we can do as scientists is to evaluate the predecessors' work critically, test new hypotheses by carefully conducting experiments, interpret results by applying the correct scientific principles, and honestly report the findings and form new theories.

This volume, *Mechanobiology of Hydrostatic Pressure*, starts with a short chapter on the physics of hydrostatic pressure and how it impacts the human body. This book also contains literature reviews to compare the effect of different mechanical stimuli, fluid flow, stretch, and hydrostatic pressure on various types of cells. Most of this volume is a collection of my previous and current research work: Previously published papers have been revised and edited for this book. In the postface, I summarized the contents of this book and briefly described the current theories on how cells may sense hydrostatic pressure. The mechanobiology of hydrostatic pressure is a growing field with many unknowns, but emerging ideas and findings continue to accumulate. I hope that the advancement of the field will help solve the mysteries of life and contribute to improving human health.

Clemson, SC, USA Jiro Nagatomi

Acknowledgments

The author would like to thank the two editors and friends, Professor Taiji Adachi of Kyoto University and Professor Hiromi Miyoshi of Tokyo Metropolitan University, for inviting me to write this volume as part of their Springer Frontiers in Biomechanics Book Series. Their suggestions to reformat the contents to include some of my previous papers helped me tremendously in completing this project. All the contributions to the work presented here are acknowledged as co-authors of the original publications and in this volume. Chapters 2.2, 2.3, 2.4, 3.2, 3.3, 3.5, 3.6, and 3.8 are adaptations of previous peerreviewed papers. Chapter 3.4 is an adaptation of the M.S. thesis by my former student, Ms. Kelly Etten. Chapter 3.8 is a contribution by my former students, Dr. Cody Dunton, Mr. Justin Bacaoat, Ms. Alexandria Pendino, and Ms. Julia Lunt. Special thanks go to all my teachers, mentors, and former students who helped shape my scientific career and my family members, both alive and deceased, who supported me through my life journey.

Clemson, SC, USA Jiro Nagatomi

Contents

Part I
Introduction

Chapter 1
Physics of Hydrostatic Pressure and Its Importance in Physiology

Contents

Hydrostatic pressure is, as the name indicates, pressure in a fluid (hydro: water in Greek) at rest (static). Pressure, by definition, is the continuous physical force exerted on or against an object by something in contact with (Simpson 1989). In the physical sciences, pressure is often defined as the perpendicular force per unit area. In this chapter, we will dissect the physics of hydrostatic pressure with examples of where hydrostatic pressure impacts nature, including biomechanics and human health. We will examine first what pressure is in a physical sense and use some mathematical equations to view it from multiple scientific and engineering disciplines.

1.1 Weather and Atmospheric Pressure

Pressure is surrounding us all. In our daily life on Earth, we encounter atmospheric pressure that constantly changes to influence the climate/weather. High-pressure systems are associated with clear skies and calm weather, while low-pressure systems are associated with clouds, precipitation, and turbulent weather like

J. Nagatomi, *Mechanobiology of Hydrostatic Pressure*, Frontiers of Biomechanics 5, https://doi.org/10.1007/978-981-95-4405-9_1

hurricanes. While direct quantitative correlations may not be available, many people agree that changes in atmospheric pressure, also called barometric pressure, affect their health differently, from headaches (Maini and Schuster 2019) to joint pain (Bongers and Vandenneucker 2020). Others have noticed that climate changes affect mental and behavioral health (Li et al. 2023). There may be something in our body that could detect small changes in atmospheric pressure.

1.1.1 Altitude and Pressure

Earth's atmospheric pressure is directly related to the altitude above sea level: the pressure decreases as the altitude increases. Our body can undoubtedly feel the altitude-dependent changes in atmospheric pressure in more than one way, indirectly or directly. We can feel the temperature drop as we travel from sea level to a higher altitude because air pressure and temperature are proportional to each other. So, our body feels the pressure decrease indirectly through temperature drops as the altitude increases. Indeed, for every 1000-meter rise in altitude, the temperature falls by 6.5 degrees Celsius. Some people might make the mistake of hiking underdressed in the mountains during summertime. Even if it is nice and warm at 25 degrees Celsius in the city, it is freezing at the top of high mountains. Another way we feel the decrease in atmospheric pressure at high altitudes is through shortness of breath. Because the density (mass per volume) of air is related to the atmospheric pressure, there is a lower amount of air (and oxygen in proportion) available for breathing than at sea level. As our body takes some time to acclimate to the difference in the amount of oxygen available for breathing, high-altitude sickness can occur when we travel from low to high altitude in a short period. Athletes like long-distance runners can take advantage of this and train at high elevations before races to increase their endurance. These examples tell us that our physiology is very sensitive to pressure changes.

1.1.2 Ocean, Sky, and Pressure

The pressure dependence on altitude and its effect on our body exists below sea level and above the mountains we can climb. Our body can feel increases in hydrostatic pressure through squeezing on our eardrums as we dive deeper into the ocean or swimming pool. The inward bowing of the eardrum and the discomfort in the ears result from the difference in pressure between the increasing pressure of water and the pressure of the air filling the middle ear that stays the same as the atmospheric pressure at sea level. Similar discomfort in the ear due to pressure inequality between the inside and outside of the eardrum may be felt during an air flight or elevator ride in a very tall building. The pain and discomfort from the squeezing on the eardrum may be alleviated by equilibrating the pressure in the Eustachian tube through simple actions like swallowing or trying to breathe out with mouth and nose closed. This is yet another example of how hydrostatic pressure changes impact our bodies.

1.2 Physics of Hydrostatic Pressure, Gas Laws and Fluid Mechanics

1.2.1 Gas Laws

Our day-to-day observations of how hydrostatic or atmospheric pressure impacts our body must be explained scientifically. Let us now dissect the physics of hydrostatic pressure and how they are described mathematically. Luckily, our predecessors have, through conducting clever experiments, come up with various physical laws for gasses (including air) that relate pressure and temperature, pressure and volume, temperature and volume, and the number of molecules and volume. All of these are combined to form the famous ideal gas law.

1.2.1.1 Boyles Law (Pressure and Volume)

As early as the seventeenth century, scientists in Britain and France were studying the pressure-volume relationship of gasses. However, Robert Boyle conducted experiments using a j-shaped glass tube and published the results to confirm that under a constant temperature, the pressure and volume of gasses are inversely proportional to each other, known as Boyle's law.

$$P \propto \frac{1}{V}$$

which can be rewritten as

$$PV = k$$

where P is the pressure of the gas, V is the volume of the gas (in Kelvins), and k is a constant.

to state that although the individual values of pressure and volume of a gas can vary, as long as the temperature remains constant, the product of these values is constant. In other words, under two different sets of conditions at constant temperature, the equation can be rewritten as

$$P_1V_1 = P_2V_2$$

This would allow the prediction of the changes in the volume of a gas under different pressures or the changes in the pressure of a gas during volume changes.

1.2.1.2 Charles Law (Volume and Temperature)

In the late eighteenth century, a French scientist/balloonist, Jacques Charles, conducted an experiment to demonstrate that the volume of gases depends on temperature by heating gas-filled balloons and observing that they all increased in volume. However, these results were never published until another scientist, Joseph Louis Gay-Lussac in 1802 confirmed and created the formula

$$\frac{V}{T} = k$$

or

$$V_1 T_1 = V_2 T_2$$

where V is the volume of the gas, T is the gas temperature (in kelvins), and k is a constant (subscripts indicate two different states of the same gas).

In fact, almost 100 years earlier, another French physicist, Guillaume Amontons, had conducted similar experiments and described this principle. Still, results were never published, and Gay-Lussac named the law in honor of Charles.

1.2.1.3 Amonton's/Gay-Lussac's Law (Pressure and Temperature)

As discussed earlier, we observed that air pressure and temperature drop with increasing altitude. Indeed, the drop in the atmospheric pressure with altitude is caused by a decrease in air molecules and water vapor due to Earth's gravity (as explained later). Moreover, many things like relative humidity, wind speed (convection), and latitude (near the equator vs. near the poles) can affect the air temperature in the atmosphere in which we exist (near the surface of the Earth). However, when the mass and volume are held constant, the gas pressure is proportional to its temperature. This pressure-temperature relationship was another formula first published by Gay-Lussac. Still, it had been observed earlier by Amontons, and the law is known as Gay-Lussac's Law or Amonton's Law.

$$\frac{P}{T} = k$$

where P is the pressure of the gas, T is the temperature of the gas (in kelvins), and k is a constant. This can also be expressed as comparing the same substance under two conditions.

$$\frac{P_1}{T_1} = \frac{P_2}{T_2}$$

or

$$P_1T_2 = P_2T_1$$

This can be explained if pressure is regarded as the force that acts on a given surface and temperature as a measure of the kinetic energy of molecules moving rapidly and randomly. As you increase the temperature, thereby the kinetic energy of a gas, its particles collide with each other and the container walls more frequently, which increases pressure, the force acting on surfaces at any given instance.

1.2.1.4 Avogadro's Law (Volume and Number of Molecules)

Using the same concept of the kinetic energy of molecules, another gas law, Avogadro's Law, may be explained. This law was named after an Italian scientist, Amedeo Avogadro, in 1812 and states that under constant temperature and pressure, the equal volume of all gases has the same number of molecules. For example, equal volumes of hydrogen and oxygen gases contain the same number of atoms at the same temperature and pressure. This can be written as:

$$\frac{V}{n} = k \text{ or } \frac{V_{hydrogen}}{n_{hydrogen}} = \frac{V_{oxygen}}{n_{oxygen.}}$$

where V is the volume, n is the number of molecules/atoms of the gas, and k is a constant for a given temperature and pressure.

1.2.2 *Ideal Gas Law*

When Boyle's, Charles', Gay-Lussac's, and Avogadro's Laws were combined by a French scientist, Cleypron, in 1834, the Ideal Gas Law was born.

$$PV = nRT$$

where R is the ideal gas constant. Rearranging this equation will give

$$\frac{V}{n} = \frac{RT}{P}$$

and comparing that to the equation for Avogadro's Law will give the constant for fixed temperature and pressure.

$$k = \frac{RT}{P}$$

Another way of describing the Ideal Gas Law uses the Boltzmann constant, k_B, and the number of particles (as opposed to the number of molecules) in the gas, N,

$$PV = Nk_BT$$

, which serves as the basis for the kinetic theory of gases that relates the motion of submicroscopic gas particles with macroscopic properties of gases such as pressure and volume.

1.2.2.1 Dalton's Law (of Partial Pressure)

Dalton's law of partial pressure is an empirical law observed by an English chemist, John Dalton, in 1802. It states that the total pressure exerted by the mixture of inert (non-reactive) gases is equal to the sum of the partial pressures of individual gases.

$$P_{total} = P_1 + P_2 + P_3 + \ldots + P_n$$

By volume, the air in the Earth's atmosphere is approximately 78% Nitrogen, 21% Oxygen, 1% Argon, and other gasses, including Carbon Dioxide making up the rest. The partial pressure of each gas follows the same ratio and adds up to the total atmospheric pressure. Dalton's Law explains why we feel out of breath at high altitudes, where the total atmospheric pressure and the partial pressure of inspired oxygen are lower than that at sea level. The lower partial pressure of oxygen means there is less force for oxygen molecules to move from the inspired air into the blood circulation through the lung. The decline in the movement of oxygen into our body, in turn, affects the physiology of the cells. This brings us to the next section, which discusses hydrostatic pressure from the concepts of fluid mechanics, which is the study of fluids (liquids and gasses) and the forces that act on them.

1.3 Hydrostatic Pressure in Fluid Mechanics

Classical fluid mechanics is based on two major ideas. One is the idea that a fluid (either a liquid or a gas) is a continuum, meaning that it is made up of infinitely divisible tiny fluid particles that, at every point, have a value for fluid properties like pressure, temperature, density, viscosity, velocity, etc. The other important idea is the application of Newton's second law of motion to these tiny particles, which is described by the famous equation,

$$\boldsymbol{F} = m\boldsymbol{a} = m\frac{dv}{dt}$$

, where F is Force, m is mass, a is acceleration, and v is velocity. This equation tells us that the rate of change in momentum (mass times velocity) for a tiny particle must balance the force that acts on the element. Two types of forces act on fluid particles: body force, like gravity, and surface force, like pressure. When you consider all the forces acting on a fluid element in three dimensions under certain assumptions like Newtonian behavior or incompressibility, the balance equations (without going through the derivation here) become the Navier-Stokes equation.

$$-\nabla p + \mu\nabla^2 \boldsymbol{v} + \rho\boldsymbol{g} = \rho\boldsymbol{a}$$

, where p is pressure, μ is viscosity, v is velocity, ρ is density, g is gravity, and a is acceleration. Furthermore, assuming the fluid is inviscid and the viscosity has negligible effects, the equation is simplified to the Euler equation.

$$-\nabla p + \rho\boldsymbol{g} = \rho\boldsymbol{a}$$

Finally, if the fluid is static, the acceleration is zero, and the equation becomes.

$$-\nabla p + \rho\boldsymbol{g} = 0$$

Since this is a vector equation, it has x-, y-, and z-components, but taking the z-axis in the vertical direction will make the gravity terms $g_x = 0$, $g_y = 0$, and $g_z = -g$ (gravitational constant) and

$$\frac{\partial p}{\partial x} = 0 \quad \frac{\partial p}{\partial y} = 0 \quad \frac{\partial p}{\partial z} = -\rho g$$

This shows that the difference in pressures at two points in the vertical direction depends on the density of the fluid and gravity. Solving the differential equation by taking a definite integral on both sides of the last equation will give

$$p - p_0 = -\rho g\left(z - z_0\right)$$

The left side is the pressure difference, and $z\text{-}z_0$ is the vertical height, h; we can see that the greater the height measured downward (e.g., depth of the ocean from the surface), the higher the pressure, and the equation can be rewritten in a more familiar form of hydrostatic pressure.

$$p = \rho g h$$

, where p is hydrostatic pressure (− atmospheric pressure, p_0), ρ is fluid density, g is gravitational constant, and h is fluid depth. As mentioned earlier in the derivation steps, there was an assumption that the fluid is incompressible (e.g., water or blood), so the actual calculation would be a little more complex for compressible fluids like air. Nonetheless, this equation also explains the relationship between the air pressure and altitude for the lower atmospheric pressure (and partial oxygen pressure) at high altitudes.

1.4 Application to Medicine

Many types of pressure inside our body are measured to assess the individual's health. The most common pressure to be measured routinely is blood pressure. Although blood moves through circulation and is not at rest, blood pressure is a measurement of hydrostatic pressure within the large artery at instances of systole (contraction) and diastole (relaxation) of the heart. A systolic pressure of 120 mmHg and a diastolic pressure of 80 mmHg are considered normal, and when the numbers are higher than these (hypertension), risks for cardiovascular disease are considered to be higher (American Heart Association 2023). In addition to the routinely measured blood pressure, one of the standard pressure measurements to assess the cardiac function of the individual is central venous pressure (CVP), which is measured inside the vena cava near the heart using a catheter. A typical range of CVP is 8–12 mmHg, and elevated values can indicate various critical conditions of the heart or lungs that require treatments (Shah and Louis 2024). Systolic pressure and CVP represent the maximum and minimum pressures in the vasculature, and this pressure difference is necessary for the blood to reach various organs and return to the heart. The blood must then enter the pulmonary circulation for gas exchange in the lungs, where airway and pleural pressures are critical respiratory function parameters. Moreover, the pressures inside the skull, eye, and abdomen, known as intracranial, intraocular, and intraabdominal pressures, are also used to diagnose various conditions.

1.5 Summary

Hydrostatic pressure, or simply pressure, is all around us in the form of atmospheric pressure that affects the weather system and inside our bodies as the driving force for moving fluids. Also, pressure is an essential physiological parameter to indicate the conditions of various organ systems. Through physical laws and mathematical equations that were established over centuries, we now understand hydrostatic pressure and its applications in science, engineering, and medicine. We can extrapolate these concepts to imagine that the cells inside our body are constantly subjected to changes in hydrostatic pressure and may respond to abnormality in a specific manner for each organ system. However, we have a limited understanding of the relationship between hydrostatic pressure and cellular events, as the mechanobiology of hydrostatic pressure has only been explored in the past few decades. Although the scope may be limited, the studies reviewed or presented in the subsequent chapters of this volume have shed some light on how cells of various organ systems behave under hydrostatic pressure stimuli.

References

American Heart Association. (2023). Understanding blood pressure readings: American Heart Association. https://www.heart.org/en/health-topics/high-blood-pressure/understanding-blood-pressure-readings. Updated 2023

Bongers J, Vandenneucker H (2020) The influence of weather conditions on osteoarthritis and joint pain after prosthetic surgery. Acta Orthop Belg 86(1):1–9. https://search.proquest.com/docview/2409192070

Li D, Zhang Y, Li X, Zhang K, Lu Y, Brown RD (2023) Climatic and meteorological exposure and mental and behavioral health: a systematic review and meta-analysis. Sci Total Environ 892:164435. https://doi.org/10.1016/j.scitotenv.2023.164435

Maini K, Schuster NM (2019) Headache and barometric pressure: a narrative review. Curr Pain Headache Rep 23(11):87–87. https://doi.org/10.1007/s11916-019-0826-5

Shah P, Louis MA (2024) Physiology, central venous pressure. In: StatPearls. StatPearls Publishing LLC, Treasure Island (FL)

Simpson JA (1989) The Oxford English dictionary. Clarendon Press, Oxford

Part II
In Vitro Studies of Fluid Flow v. Hydrostatic Pressure

Chapter 2
Cellular Responses to Fluid Flow Vs. Hydrostatic Pressure

Contents

J. Nagatomi, *Mechanobiology of Hydrostatic Pressure*, Frontiers of Biomechanics 5, https://doi.org/10.1007/978-981-95-4405-9_2

2.1 Cellular Responses to Fluid Flow Vs. Hydrostatic Pressure

As described in Chap. 1, hydrostatic pressure (HP) exists within a fluid at rest in equilibrium and any difference in HP between two points in space would drive the fluid to flow along the pressure gradient from the higher to lower pressure point. For this reason, many types of cells in different organs including those of the cardiovascular and skeletal systems are constantly subjected to both fluid flow (and associated shear stress) and HP simultaneously. Since the 1980's the advancement in cell cultures allowed investigators in the field of fluid mechanics to study the link between fluid flow-induced shear stress and vascular health. Moreover, similar experimental approaches have been utilized to expose bone cell cultures to fluid flow and study the century-old theory that the skeletal tissues adapt to mechanical loading, known as Wolff's Law. To a lesser extent, investigators also examined the effects of physiological level HP on these cells. In this chapter, we will review the anatomy, physiology, and cellular composition to explore the biomechanical environment of the cells of these organs. We will then review the in vitro studies that aim to understand the role of flow-induced shear stress and HP in pathophysiology of vascular and skeletal organs.

2.1.1 Anatomy, Physiology, and Mechanical Loading Environments for Vascular Cells In Vivo

At the center of the cardiovascular system, the heart is connected to the systemic circulation to supply oxygenated blood for the entire body, and to the pulmonary circulation that gets blood oxygenated in the lungs. The systemic circulation consists of the left side of the heart, the aorta, large and small arteries, which transport blood to various organs through capillaries, and small and large veins that carry blood back to the right side of the heart through the venae cavae (McKinley and O'Loughlin 2005). The pulmonary circulation consists of the right side of the heart, the pulmonary artery which branches into smaller arteries, arterioles, and lung capillaries where gas exchange occurs, and the pulmonary veins carry the oxygenated blood back to the left side of the heart. All arteries and veins are made up of three layers, tunica intima, tunica media, and tunica adventitia with different thicknesses. The tunica intima and lumen of the capillaries are lined with endothelial cells while the tunica media consists of smooth muscle cells surrounded by the connective tissue (collagen and elastin) of the tunica adventitia (McKinley and O'Loughlin 2005).

The circulation of blood is achieved by the pumping action of the heart, which generates pressure as high as 100 mmHg on average. Since the pumping of the heart is pulsatile, the pressure in the large arteries fluctuates between 120 mmHg when it contracts and 80 mmHg when it relaxes, commonly described as the top and bottom blood pressure. As the blood is transported through the systemic circulation, the pressure drops progressively to almost 0 mmHg when it reaches the right side of the heart, with the functional pressure at 17 mmHg on average in the systemic capillaries (Hall and Guyton 2016). This pressure difference across the systemic circulation drives the blood to flow through the arteries and veins. As a result, the endothelial cells that line the blood vessel walls are constantly subjected to shear stress induced by the tangential force of the flowing blood acting over the surface, in addition to being subjected to varying levels of HP acting perpendicular to the cell surface. Fluid flow-induced shear stress depends on the velocity and other characteristics of the blood flow and ranges between 10–70 dyn/cm^2 in the arteries and 1–6 dyn/cm^2 in veins (Papaioannou et al. 2006). High values are observed in the regions with high velocity blood flow and low shear stresses are associated anatomical features (e.g., arterial bifurcations) that promote turbulent and unstable flows (Papaioannou et al. 2006). While endothelial cells are constantly subjected to normal, physiological levels of HP and shear stresses, which are necessary for maintenance of vascular health, abnormal changes in hemodynamic parameters have been associated with various pathological conditions such as atherosclerosis and aneurysms (Papaioannou et al. 2006). Studies have shown that unidirectional steady laminar flow and high shear stress stimulate atheroprotective endothelial cell functions including maintenance of barrier function and suppression of blood coagulation and immune cell adhesion (Reinhart-King et al. 2008). In contrast, disturbed flow and low shear stress stimulate endothelial cell phenotypic shifts and loss of function that make them prone to plaque formation (Chiu et al. 2009).

2.1.2 *Effects of Shear Stress on the Cells of the Vascular System*

To elucidate the underlying mechanisms for shear stress effects on vascular diseases such as atherosclerosis, numerous in vitro studies have examined various endothelial and smooth muscle cell responses to controlled steady and unsteady flow conditions. There have been two general approaches to apply flow-induced shear stress on cultured cells, namely, cone-and-plate viscometer type and parallel plate flow chamber type systems with the latter being more common lately. Both systems are based on the theory of Newtonian fluid flow in which the shear stress on the plate is determined by the equation

$$\tau = \mu \frac{du}{dy} \tag{2.1}$$

where τ: shear stress, μ: kinematic viscosity, u: fluid velocity, y: the distance from the plate surface, du/dy: velocity gradient (shear rate). The precise geometry of the cone-and-plate, or parallel plate produces a uniform velocity profile to allow application of uniform shear stress stimuli to the cells cultured on the bottom plate.

One of the earliest of these studies used a cone-plate apparatus to expose cultured bovine aortic endothelial cells to a laminar shear stress of 5–10 dynes/cm^2, and demonstrated that confluent monolayers undergo a time-dependent change in cell shape to become uniformly oriented with flow (Dewey et al. 1981). Moreover, another team reported that the endothelial cells morphological shift was dependent on the magnitude and direction of shear stresses (Levesque and Nerem 1985). Over the past four decades since these seminal studies were published, the advancement in cell and molecular biology approaches allowed elucidation of the relationship between shear stress and endothelial function as well as the potential sensing mechanisms. For example, shear stress has been shown to regulate endothelial nitric oxide (NO) synthase (eNOS), an enzyme that generates NO, which, in turn, regulates endothelial cell survival, vasodilation, angiogenesis and provides anti-inflammatory and antioxidant properties (Chistiakov et al. 2017). The shear stress-induced eNOS up-regulation is mediated by activation of a transcription factor, KLF2 (Dekker et al. 2002), which also plays a role in maintaining endothelial homeostasis through increasing anti-thrombotic properties (Lin et al. 2005) and inhibiting expression of proinflammatory and adhesive molecules (Atkins and Jain 2007). The in vitro studies using flow systems have also shown that these mechanically regulated events in endothelial cells are mediated by multiple mechanosensing molecules. To date, proposed endothelial cell mechanosensors include, but are not limited to, cell-cell junctional proteins, integrins, focal adhesions, ion channels, receptor kinases, G-protein coupled receptors, cytoskeletal proteins, and glycocalyx (Chistiakov et al. 2017; Conway and Schwartz 2012)). Among these mechanosensing molecules, the most well-studied and endothelial specific may be the mechanosensory complex that consists of PECAM-1 (platelet endothelial cell adhesion molecule 1), VE-cadherin (vascular-endothelial cadherin) and VEGFR2 (vascular endothelial growth factor receptor 2) at endothelial cell-cell junctions (Conway and Schwartz 2012).

Studies have shown that the effects of shear stress on endothelial cells are communicated with the underlining vascular smooth muscle cells or vice versa (Chistiakov et al. 2017). For example, in a co-culture model of endothelial cells and smooth muscle cells, exposure to laminar shear stress led to suppression of cell migration (Sakamoto et al. 2006) and maintenance of contractile phenotype in smooth muscle cells (Sakamoto et al. 2011; Tsai et al. 2009). In contrast, exposure of a co-culture or endothelial and smooth muscle cells to disturbed flow resulted in differentiation of smooth muscle cells to synthetic phenotype by reduction in a-smooth muscle actin, SM22, and smooth muscle myosin heavy chain and proinflammatory phenotype by induction of VCAM-1, IL-8, and monocyte chemoattractant protein-1 genes (Hastings et al. 2007). The shear-induced crosstalk between endothelial cells and smooth muscle cells is mediated, at least in part, by soluble factors released by endothelial cells such as PDGF-BB and TGF-β1 through paracrine and feedback mechanisms (Qi et al. 2011). In addition, endothelial cells have been shown to release exosomes containing miRNAs in response to shear stress, which are taken up by smooth muscle cells for regulation of cell differentiation (Hergenreider et al. 2012). The data accumulated through these and numerous other studies from the past four decades have provided significant knowledge on effects of shear stress on vascular health. However, as described above, vascular cells are subjected to not only flow-induced shear stress, but also to HP through blood pressure, which has been shown to impact vascular health and pathology.

2.1.3 *Effect of HP on Vascular Cells*

Compared to the volume of knowledge on the effect of shear stress on vascular endothelial cells and smooth muscle cells, relatively little is known about the responses of these cells to HP. Nonetheless, the results of in vitro studies in the past three decades have demonstrated that cell morphology, cell proliferation, cell migration, and apoptosis of endothelial cells are influenced by elevated HP in a wide range (from 3 mmHg to 150 mmHg) (Acevedo et al. 1993; Schwartz et al. 1999; Salwen et al. 1998; Shin et al. 2002). In general, two approaches have been used to apply elevated HP on cultured cells, hydrostatic column-based or pressurized gas-based. The hydrostatic column-based approach is based on the principle that the HP value is directly proportional to the depth of fluid.

$$P = \rho g h \tag{2.2}$$

Where ρ is the density of the fluid, g is the gravitational constant, and h is the height. By placing adherent cells seeded on culture substrates such as a glass coverslip at the bottom of a container connected to a column of cell culture buffer solution, HP that is proportional to the height of the column is applied to those cells. While the advantage of this approach is the ease of creating a hydrostatic head, relative height difference between the cultured cells and fluid surface, application of high pressure (to simulate high blood pressure ~ 200 mmHg) is physically challenging. In contrast, the pressurized gas-phase approach is based on the principle that the HP at the bottom of fluid-filled container is the sum of the HP due to the height of the fluid and the pressure of the gas phase above the fluid (P_o).

$$P = \rho g h + P_o \tag{2.3}$$

By placing the cell culture container inside another sealed chamber for pressurization of humidified gas mixture, application of higher pressure to cultured cells can be achieved.

One of the earliest studies used the hydrostatic column-based approach to expose sub-confluent monolayer of arterial endothelial cells to HP in the order of 10–15 cmH_2O and demonstrated in response to elevated HP the cell proliferation increased and cell morphology changed to form multiple layers, which were mediated by basic fibroblast growth factor (Acevedo et al. 1993; Salwen et al. 1998) . In a separate study, venous endothelial cells proliferation was also shown to be stimulated by exposure to elevated HP (4 cmH_2O) after 5 days and involvement of focal adhesion complex in HP mechanotransduction was observed (Schwartz et al. 1999). In studies that examined the effects of HP in more physiological ranges 50–150 mmHg, endothelial cells exhibited not only increased cell proliferation, but changes in the barrier function through altered expression of intracellular junction proteins such VE-cadherin (Ohashi et al. 2007) and zona occludens (ZO)-1 (Shin et al. 2003) as well as reorganization of cytoskeletal filaments (Ohashi et al. 2007; Shin et al. 2003). These results provided evidence that endothelial cells are sensitive to elevations in HP as small as a few mmHg, indicating that elevation in blood pressure alone can influence endothelial function and vascular health.

In addition to the effect on vascular pathology, studies have shown that HP can promote vascular tubulogenesis, which is a process required for the formation of vascular network under normal physiological conditions (Yoshino et al. 2020; Shin et al. 2012). For example, 3D culture of bovine aortic endothelial cells exposed to elevated HP (20 mmHg) for 3 days exhibited significantly upregulated sprout formation compared to controls, and the response was enhanced in the presence of exogenous fibroblast growth factor-2 (FGF-2) (Shin et al. 2012). In a separate study, human umbilical vein endothelial cells in a 2D culture exhibited an increase in formation of tube-like structures when exposed to HP (50 mmHg) for 3 hrs, and this pressure-induced tubulogenesis was mediated by activation of the Ras-ERK signaling pathway even in the absence of FGF (Yoshino et al. 2020) (Yoshino 2020). Regardless of the mechanisms, growth factor mediated (Shin et al. 2012) or G-protein coupled receptor-mediated (Yoshino et al. 2020), the discovery that vascular tubulogenesis can be promoted by elevated HP provided insight into understanding tumor angiogenesis induced by interstitial pressure, as well as optimization for vascularization of engineering tissues in vitro.

Despite the number of discoveries made on the endothelial cellular and molecular responses to HP, the exact mechanisms for how endothelial (or any other mammalian) cells sense pressure are still not well understood. Among the various structural elements of endothelial cells, the most promising candidates of pressure sensors are the membrane-bound channels (Yoshino et al. 2020; Friedrich et al. 2019; Guo et al. 2016; Prystopiuk et al. 2018) More specifically, involvement of cation channels such as endothelial sodium channels (ENaC) (Prystopiuk et al. 2018), transient receptor potential (TRP) channels (Prystopiuk et al. 2018), and piezo-1 channel (Friedrich et al. 2019), as well as aquaporin (AQP) water channel (Yoshino et al. 2020) has been implicated in pressure sensing by endothelial cells. The reasons that these membrane-bound ion channels have received as potential pressure sensors are that endothelial cells exhibit increases in the intracellular calcium concentration under elevated HP and many of the signaling events that lead to observed changes require increases in intracellular calcium ions (Nilius and Droogmans 2001). For example, when exposed to elevated HP as high as 100 mmHg the intracellular calcium oscillation and concentration in aortic endothelial cells significantly increased compared to the baseline level in the control group (Prystopiuk et al. 2018; Kuebler et al. 2002). In the presence of mechano-sensitive channel blockers such as gadolinium and GsMTx-4, or ENaC blocker amiloride, however, the intracellular calcium concentration of endothelial cells remained at the baseline level, indicating the involvement of these channels in the endothelial cell calcium response to HP (Prystopiuk et al. 2018; Kuebler et al. 2002). Moreover, it was found that the retraction velocities of cytoskeletal filaments in endothelial cells increased under elevated HP, but this pressure-dependent change was abolished in the presence of those channel blockers that inhibited calcium increases (Prystopiuk et al. 2018). Interestingly, another more recent study demonstrated that exposure of human umbilical vein endothelial cells to elevated HP did not induce an increase in the intracellular calcium concentration, but resulted in efflux of a fluorescent calcium indicator from the cytoplasm (Yoshino et al. 2020). Since this efflux of the calcium indicator occurred concomitantly with cell contraction, and is was inhibited by blocking AQPs, the authors concluded that HP is sensed by endothelial cells through water channels, rather than ion channels (Yoshino et al. 2020). Although it is possible that the endothelial cells from the high-pressure vessels and low-pressure vessels function differently, the exact sensing mechanism for HP remains controversial.

2.1.4 Anatomy, Physiology, and Mechanical Loading Environments for Bone Cells In Vivo

The skeletal system consists of the bones of the skeleton, cartilage, ligaments, and other connective tissues that keep the skeleton stable. The main component of the bones is the mineralized bone tissue matrix, but they also contain periosteum, articular cartilage, bone marrow, blood vessels, and nerves. The bones support the body weight, protect vital organs, and interact with muscles to achieve various highly controlled motions of body segments. In addition, the bones are the reservoirs of minerals such as calcium and phosphorus, and blood cells are produced within bone marrow. Four major types of cells are found within the connective tissue of the bone: mesenchymal stem cells, osteoblasts, osteocytes, and osteoclasts. Mesenchymal stems cells are found in bone marrow and have the potential to differentiate into multiple types of cells that form various tissues including bone, cartilage, fat, muscle, nerve, liver, blood vessels, etc. Osteoblasts are formed from mesenchymal stem cells and secrete the organic components of the bone matrix such as collagen and other bone-related proteins (osteocalcin, osteopontin, osteonectin, etc.) that is later mineralized to form the hard bone tissue. Osteoblasts are found on bony surfaces and they can be either actively synthesizing new bone matrix or inactive. But, these inactive osteoblasts, also known as bone-lining cells, can become activated to form new bone when needed. Once osteoblasts complete formation of new bone, they differentiate into osteocytes and are entrapped in small spaces called lacunae within the mineralized matrix. Osteocytes maintain intracellular contact and communication with numerous processes that extend through interconnecting channels called canaliculi, which also serve as conduits for water and nutrient transport.

It has been recognized for over a century that a relationship exists between mechanical loading on the skeleton and bone structure (Turner 1992) . Reduction in skeletal loading due, for example, to prolonged bedrest or spaceflight has been associated with bone loss (Man et al. 2022). In contrast, bone mass has been shown to increase in response to increased loading through exercise (Turner and Robling 2003). These observations have been reproduced in controlled animal model experiments to artificially create either disuse of limbs (Bateman et al. 2000) or increases in skeletal loading (Torrance et al. 1994). Indeed, unloading of limbs in mice models led to decreased bone mass, which can be mitigated by inhibition of osteoclast function (Bateman et al. 2000) and dynamic axial loading on the ulna of rats and mice led to increased bone mass (Torrance et al. 1994; Robling et al. 2002). While these studies demonstrated that relations exist between mechanical loading and bone cell function, it is not clear whether osteoclast and osteoblast are directly activated by unloading and loading on the bone. Mechanically speaking, the mineralized bone matrix is so rigid and physiological loading can generate only up to 1000 microstrains (Lanyon et al. 1975). This level of deformation may or may not be large enough to elicit responses cells to affect bone mass. For this reason, it was hypothesized that physical deformations of bone tissue create HP gradients within the lacunar-canalicular network where osteocytes and their processes reside under interstitial fluid (Han et al. 2004). Moreover, the trabecular bone on the ends of long bones (e.g. femurs) is highly vascularized and the difference between the arterial pressure for entry of oxygenated blood and the venous pressure at the exit generates so-called intraosseous pressure within the bone marrow. Thus, compression on the marrow filled trabecular bone increases the intraosseous pressure (Beverly et al. 2020), which, in turn, may create HP gradients between the marrow cavity and outside layer of the bone (Bergula et al. 1999). Regardless of the sources, those pressure gradients presumably move interstitial fluid from regions of higher pressure to regions of lower pressure, and, thus, osteocytes are exposed to fluid flow-induced shear stresses. This theory of increased fluid flow around osteocytes under skeletal loading was then supported experimentally in animal models (KnoTate et al. 2000; Price et al. 2011), and stimulated a number of in vitro studies that examined the effect of shear stress on bone cells (Reich et al. 1990; Jacobs et al. 1998; Tanaka et al. 2005; McGarry et al. 2005; Bancroft et al. 2002).

2.1.5 *Effects of Shear Stress on Bone Cells In Vitro*

In vitro models to investigate the effects of fluid flow-induced shear stress on bone cells may be categorized by the types of experimental setup used and by the types of cells studied. For example, parallel-plate flow chamber systems that are frequently used to study the effect of shear stress on endothelial cells have been also used in many studies with bone cells (Wittkowske et al. 2016). This is because they allow application of controlled flow over monolayer of cells and the shear stresses experienced by the cells can be easily calculated based on the flow rate (Q), viscosity of the media (μ), and geometry of the chambers (b: width and h: height).

$$\tau = \frac{6Q\mu}{bh^2} \tag{2.4}$$

Connecting the flow chamber to various pump systems allow application of steady (Reich et al. 1990), pulsatile and oscillatory (Jacobs et al. 1998) flow profiles with known shear stresses on the cells of interest. Another approach to investigate the effects of flow-induced shear stress on bone cell function is to use 3D culture systems in a perfusion bioreactor (Bancroft et al. 2002). This type of studies is primarily motivated by developing technology for tissue engineering applications, especially to understand the long-term effects of fluid flow on cell differentiation and mineralization of engineered 3D tissue constructs. Therefore, bone cells are often seeded on porous scaffolds made of ceramics and polymers that mimic the architecture and properties of natural bone. While the use of 3D cultures may provide a more physiological cell environment compared to rigid flat surfaces used in parallel-plate flow chambers, because of the complex pore geometry of the scaffolds, estimating shear stress experienced by cells can become challenging.

Although the original and current theory for bone cell responses to fluid flow considers osteocytes as the primary type of cells that sense the flow-induced shear stress because of their abundance and location within bone matrix, a number of studies, especially the early ones, examined the effect of shear stress on osteoblasts. This is partly because osteoblast primary cultures and osteoblast-like cell lines were widely available prior to establishment of osteocyte-like cell lines and some considered that osteoblasts might also respond to low level of flow-induced shear stress in vivo (McGarry et al. 2005). For example, osteoblasts have been shown to respond, in seconds to minutes, to relatively high levels of fluid shear stress (~2 Pa) by increasing intracellular calcium (Hung et al. 1995; You et al. 2001), inositol triphosphate (Reich et al. 1990), prostaglandin E2 (Reich et al. 1990), and Nitric Oxide (Johnson et al. 1996). Moreover, fluid flow has been shown to regulate osteoblast gene expression for osteopontin (You et al. 2001), COX-2, and c-FOS (Pavalko et al. 1998). Under low shear stresses (< 0.5 Pa), osteoblasts also increase intracellular calcium (Kou et al. 2011), collagen I mRNA (Xing et al. 2014; Yu et al. 2014), and other osteogenic marker genes such as Runx2, ALP, and osteocalcin (Yu et al. 2014).

As mentioned above, osteocytes are mature osteoblasts that are embedded within the mineralized bone matrix and comprise over 90% of bone cells (Wittkowske et al. 2016). Because osteocytes form networks with neighboring cells through canaliculi and fluid transport occurs through those channels, they are considered the primary cells that sense mechanical loading on the skeleton via fluid flow and regulate bone homeostasis. Osteocytes themselves do not actively resorb old bone or form new bone matrix, and, thus, investigators focused on fluid flow effects on intracellular signaling between osteocytes and osteoclasts or osteoblasts. For example, when an osteocyte-like cell line, MLO-Y4 cells were exposed to physiological shear stress using a parallel plate flow chamber, these cells exhibited lengthening of the dendritic processes and increased expression of a gap junction protein (Cheng et al. 2001) as well as increased release of ATP through connexin hemichannels (Genetos et al. 2007). Moreover, exposure of osteocyte-like cells to physiological fluid flow led to changes in expression of RANKL and OPG that are important for osteoclast regulation (Li et al. 2011; Li et al. 2013) and sclerostin, a regulator of osteoblast bone formation (Li et al. 2013; Spatz et al. 2015).

In addition to studies that investigated the effect of fluid flow on osteoblasts and osteocytes for its role in bone homeostasis, much research has been conducted on the effect of flow-induced shear stresses on bone marrow cells in 3D cultures (Bancroft et al. 2002; Gandhi et al. 2019; McCoy et al. 2013; Sikavitsas et al. 2005). These studies were motivated by a hypothesis that mechanical loading on the skeleton can generate local pressure gradients within the marrow cavity to drive the interstitial fluid flow and the progenitor cells of osteoblasts may respond to the shear stress stimuli (Gurkan and Akkus 2008). The concept of flow-induced bone remodeling has also been employed for stimulating cellular function in bone tissue engineering applications (Bancroft et al. 2002; Sikavitsas et al. 2005). For example, when rat mesenchymal stem cells seeded on 3D porous scaffolds were cultured under flow perfusion they deposited significantly higher amounts of mineralized extracellular matrix compared to those maintained under static conditions (Sikavitsas et al. 2005), and the responses were dependent on the flow rate (Bancroft et al. 2002) indicating that the cells were sensitive to shear stress magnitude. It was also shown that mesenchymal stem cells cultured in a perfusion bioreactor exhibited higher vascularization and mineralized tissue formation when implanted in a rat cranial defect regeneration model (Gandhi et al. 2019). These and the results of other studies provided evidence that fluid flow-induced shear stress is an important stimulus that triggers responses in osteocytes, osteoblasts, and bone marrow cells for maintenance of healthy bone mass.

2.1.6 *Effects of HP on Bone Cells*

As described above, mechanical loading on the skeleton results in increases in pressure within the marrow cavity (Beverly et al. 2020) and the flow of interstitial fluid in the lacuna-canalicular network is driven by pressure gradients (KnoTate et al. 2000). The elevation in the pressure of the bone marrow and lacuna, therefore, must be experienced by the cells such as mesenchymal stem cells, osteoblasts, and osteocytes, and HP may serve as an additional triggering stimulus for these cells to induce responses.

In vitro studies that examined the effect of HP on bone cells have been reported since the 1990s (Ozawa et al. 1990; Brighton et al. 1996; Klein-Nulend et al. 1995) and these early studies established the approach that is still used in recent studies (Zhao et al. 2015; Kitamura et al. 2017), to increase the HP experienced by the monolayer cells at the bottom of culture vessels by pressurization of gas phase above the cell culture media. Variations of these custom pressure systems typically use compressed gas mixture as a pressure source and electronic components such as solenoid valves to generate cyclically oscillating HP stimuli inside a pressure chamber. Another common approach to achieve exposure of cultured bone cells to elevated HP is direct pressurization of culture media within a closed system (Stavenschi et al. 2018; Gardinier et al. 2009). Similar to parallel plate flow chambers that are used to expose cells to shear stress, these direct pressure systems often use a syringe pump connected to a small chamber containing cell-seeded glass slides. Because the fluid-filled system is closed and the culture media is incompressible, cyclic motion of the syringe piston generates cyclically oscillating HP stimuli without generating shear stress due to fluid motion. The effect of HP on bone cells can be examined not only on monolayers, but also 3D cultures using cell-seeded trabecular bone constructs in a pressure chamber (Takai et al. 2004), or hydroxyapatite / tricalcium phosphate bead cultures with a perfusion bioreactor (Aw Yong et al. 2022).

As it was the case for many studies that examined the effect of shear stress on bone cells, the early studies that examined the effect of HP on bone cells used calvaria-derived primary osteoblasts (Brighton et al. 1996; Klein-Nulend et al. 1995; Roelofsen et al. 1995; Nagatomi et al. 2001) and osteoblast cell lines (MC3T3-E1) (Ozawa et al. 1990; Klein-Nulend et al. 1997). Although the differences in the experimental conditions from different studies make it difficult to make direct comparisons of the results, reviewing of these studies suggest a general trend that osteoblast responses to HP depend on experimental parameters such as the frequency (of cyclic HP, or continuous vs cyclic), duration, and number of cycles. For example, it was shown that continuous application of HP (2 atm ~200 kPa) led to inhibition (Ozawa et al. 1990), by cyclic or intermittent application of HP (up to 40 kPa) led to increases (Roelofsen et al. 1995; Nagatomi et al. 2001) in alkaline phosphatase activity. Moreover, exposure of osteoblasts to a very short duration (1–10 s) of cyclic HP led to an increase of cell proliferation, which was similar to control (Brighton et al. 1996; Nagatomi et al. 2001) or decreased (Nagatomi et al. 2001) under longer exposure periods of 1–20 min or 1 h, respectively. With the increased availability of different cell lines, several studies demonstrated that osteocytes responded to cyclic HP by enhanced cell viability (Takai et al. 2004), increases in intracellular calcium, altered microtubule organization, increases in COX-2 mRNA level, and decreased apoptosis (Liu et al. 2010). In addition, exposure of osteocytes in 3D cultures to elevated HP via a perfusion bioreactor demonstrated HP's effect on bone remodeling gene expression (Aw Yong et al. 2022). These findings are complemented by other studies that demonstrated the effect of both cyclic and continuous HP to induce osteogenic differentiation of bone marrow stem cells (Kitamura et al. 2017; Stavenschi et al. 2018), which is mediated, at least in part, by activation of PIEZO1 mechanosensitive ion channels by HP (Kitamura et al. 2017). Taken together, HP has been shown to be an important mechanical stimulus for bone cells, namely osteoblasts, osteocytes, and bone marrow cells, to maintain bone homeostasis.

2.2 Parameter-Dependent Responses of Osteoblasts to Cyclic Pressure

2.2.1 Introduction

Initial developments in bone tissue engineering demonstrated that osteoblasts could adhere and grow on polymer scaffold surfaces. However, inducing uniform bone formation throughout the scaffold's three-dimensional structure remained challenging (Ishaug et al. 1997). While these early studies addressed key material considerations such as biocompatibility, architecture, and composition, they were conducted under static (non-mechanical) conditions. Since then, many researchers have speculated that the absence of physiological mechanical stimuli may hinder the development of engineered bone tissue in vitro and have introduced a variety of bioreactor designs to simulate mechanical forces (refer to Sect. 2.1).

In the living body, bone exists in a dynamic setting where osteoblasts, osteoclasts, and osteocytes are influenced by various physical forces. Interstitial fluid pressure, as reviewed in Sect. 2.1, is considered a significant mediator linking skeletal mechanical loads to cellular responses. For instance, rats undergoing regular physical activity exhibited increased bone formation in gravity-affected regions such as the distal tibia, compared to less-pressurized areas like the proximal tibia or lumbar spine (Iwamoto et al. 1999). Conversely, extended bed rest in humans, which removes gravitational pressure gradients, led to bone loss in the lower body but gain in the skull (Leblanc et al. 1990). While such findings emphasize the role of pressure in bone physiology, they do not explain the cellular mechanisms involved. In the present study, a custom-built system was employed to apply cyclic pressure with varying magnitudes, frequencies, and exposure durations to cultured osteoblasts. The goal was to investigate how these mechanical conditions affect proliferation, gene expression of alkaline phosphatase and osteopontin, and enzyme activity.

2.2.2 Summary of Experimental Results

Osteoblasts were harvested from neonatal Sprague-Dawley rat calvaria (Puleo et al. 1991) and cultured in DMEM supplemented with 10% fetal bovine serum under standard laboratory conditions (37°C, humidified 5% CO2). A total of 42,000 cells were seeded on treated glass coverslips and exposed to cyclic pressure ranging from 10 to 40 kPa at either 0.25 Hz or 1.0 Hz for 1 h daily over a span of up to 5 days, using a computer-controlled cyclic pressure apparatus (see Fig. 2.1). Control groups were cultured identically but without pressure exposure.After 1-h cyclic pressure exposure, the pH and pCO2 in the culture medium were similar to unpressurized controls. However, pO_2 levels increased by 15–20%. Media values were consistent

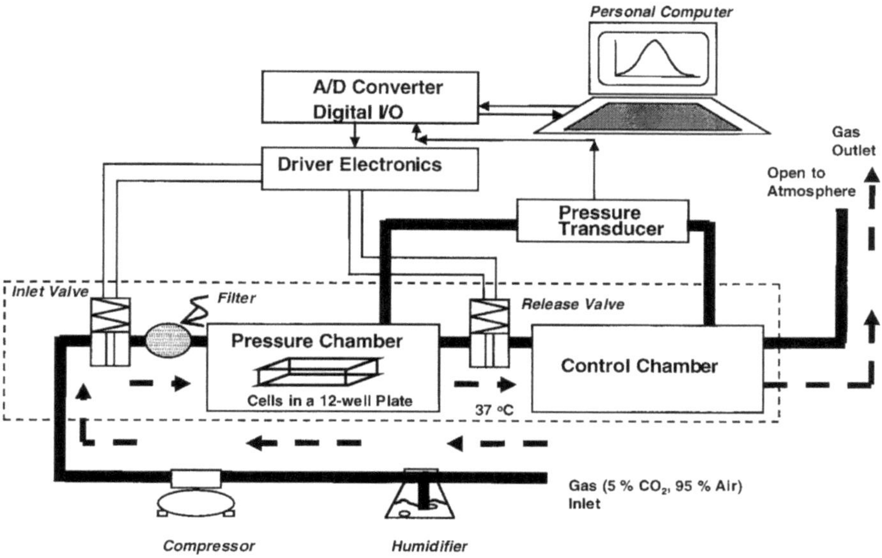

Fig. 2.1 Illustration depicting the computer-operated cyclic pressure system used in this study. A custom-made, cyclic pressure system was developed, constructed, and employed to expose cultured cells to controlled cyclic pressure regimes under standard cell culture conditions. The pressure of the gas phase above the culture medium was monitored via a pressure transducer and controlled by operating inlet and release solenoid valves using computer software specially written for this purpose. Source: Modified from Tissue Engineering, Mary Ann Liebert (2001)

with freshly prepared DMEM under ambient conditions, indicating that the observed effects were due to pressure rather than media changes.

Cell viability remained high (>90%) across all groups. No detachment was observed. When pressure was applied for just 20 min per day for 6 days, proliferation remained similar to control levels (Fig. 2.2). Likewise, no significant differences in proliferation were noted after one or 3 days of 1-h pressure exposure (Fig. 2.3). However, when cells experienced 1.0 Hz pressure for 1 h daily over 5 days, proliferation significantly declined compared to both controls and the 0.25 Hz group ($p > 0.025$).

Gene expression was examined via RT-PCR. Osteopontin mRNA levels were unchanged across all test conditions and durations (Fig. 2.4). In contrast, alkaline phosphatase mRNA was undetectable on day one but appeared after 5 days. Notably, expression was highest in cells exposed to 1.0 Hz pressure for 1 h daily, surpassing both the control and 0.25 Hz groups.

To validate gene expression changes, enzyme activity of alkaline phosphatase was measured (Thompson and Puleo 1996) and normalized to protein content and cell number. After 5 days of 1-h daily exposure at 1.0 Hz, activity was significantly elevated compared to both controls and the 0.25 Hz group ($p > 0.05$) (Fig. 2.5).

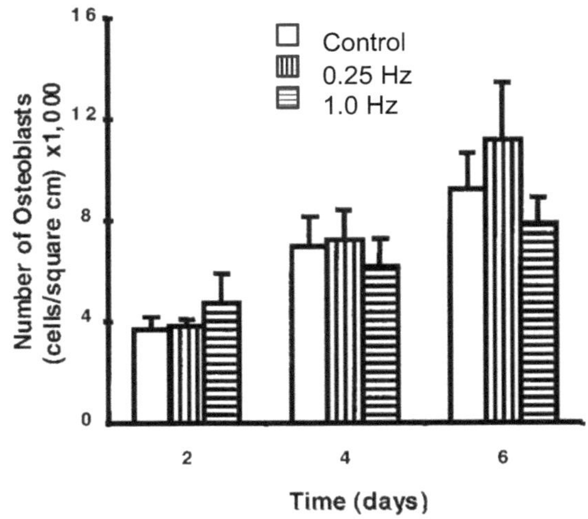

Fig. 2.2 Exposure to cyclic pressure does not affect osteoblast proliferation following 20-min exposure. Osteoblast proliferation was similar when osteoblasts underwent exposure to pressure for 20 min daily for 2, 4, and 6 consecutive days. Values are mean ± SEM; n = 3. Source: Modified from Tissue Engineering, Mary Ann Liebert (2001)

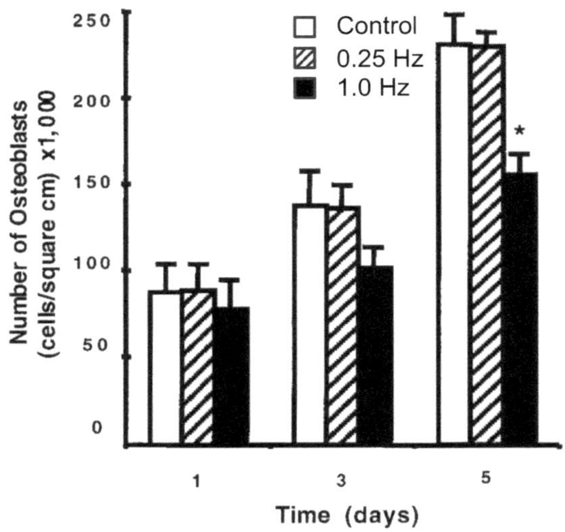

Fig. 2.3 Exposure to cyclic pressure results in decreased osteoblast proliferation in a frequency-dependent manner. Compared to controls and to osteoblasts exposed to cyclic pressure of 0.25-Hz frequency, osteoblast proliferation was notably ($p > 0.025$) decreased when these cells underwent exposure to cyclic pressure at 1.0-Hz frequency for 5 consecutive days. Values are mean ± SEM; n = 3; *$p > 0.025$ (ANOVA and Dunnett's test) Source: Modified from Tissue Engineering, Mary Ann Liebert (2001)

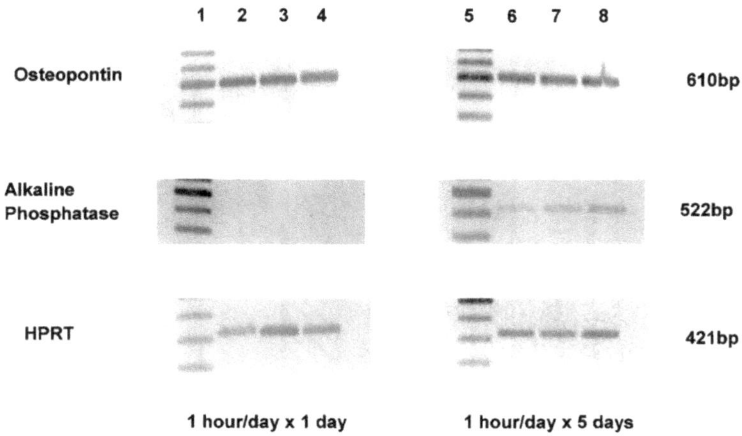

Fig. 2.4 Cyclic pressure induces differential gene expression for alkaline phosphatase and for osteopontin by osteoblasts. RT-PCR amplification of osteoblast RNA (1 mg) showed that expression of the osteopontin gene was similar under all conditions of interest to, and for all times tested in, the present study (osteopontin, lanes 2–4 and 6–8). In addition, the alkaline phosphatase gene was not expressed under all conditions tested in the present study at 1 day (alkaline phosphatase, lanes 2–4). In contrast, compared to osteoblasts maintained under control (no pressure) conditions (alkaline phosphatase, lane 6) and to osteoblasts exposed to cyclic pressure at 0.25 Hz (alkaline phosphatase, lane 7), enhanced expression of alkaline phosphatase was observed when osteoblasts underwent exposure to cyclic pressure at 1.0-Hz frequency for 1 h daily for 5 consecutive days (alkaline phosphatase, lane 8). Lane 1, base pair markers; lane 2, control (no cyclic pressure), 1 day; lane 3, 0.25 Hz, 1 day; lane 4, 1.0 Hz, 1 day; lane 5, base pair markers; lane 6, control (no cyclic pressure), 5 days; lane 7, 0.25 Hz, 5 days; lane 8, 1.0 Hz, 5 days. Source: Modified from Tissue Engineering, Mary Ann Liebert (2001)

2.2.3 *Discussion*

This investigation revealed specific relationships between cyclic pressure parameters and osteoblast responses. The selected pressure range (10–40 kPa) aligns with physiological intraosseous pressures reported during activities such as walking. For example, in humans, femoral head pressure can reach approximately 28 kPa during gait cycles (Downey et al. 1988), supporting the experimental design.

A key finding was that osteoblast proliferation is sensitive to both the duration and frequency of pressure application. Decreased proliferation occurred only with 1-h daily exposure at 1.0 Hz over 5 days, not with shorter durations or lower frequencies. These results echo in vivo findings where bone formation increased only above a stimulation frequency threshold of 0.5 Hz (Turner et al. 1994). Therefore, both frequency and the total number of pressure cycles (e.g., 3600 cycles for 1 Hz over 1 h) may determine cellular responsiveness.

Additionally, the study demonstrated that expression and activity of alkaline phosphatase, a critical enzyme for bone mineralization, were enhanced by 1.0 Hz pressure application. This suggests that cyclic pressure at higher frequencies can

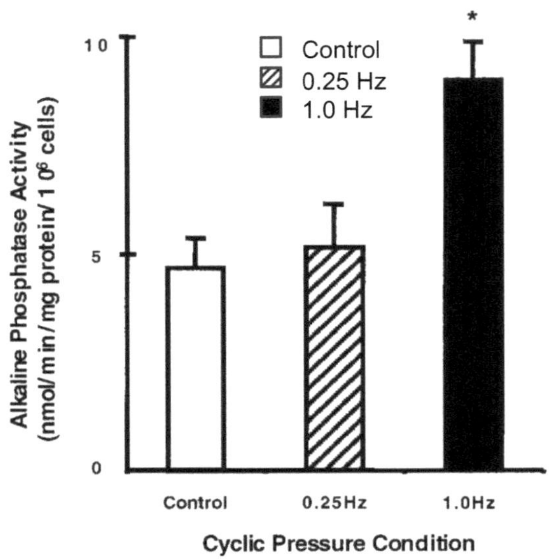

Fig. 2.5 Alkaline phosphatase activity is increased when osteoblasts are exposed to cyclic pressure at 1.0-Hz frequency. Compared to controls and to cells exposed to cyclic pressure at 0.25 Hz, alkaline phosphatase activity notably ($p > 0.05$) in- creased when osteoblasts underwent exposure to cyclic pressure at 1.0-Hz frequency for 1 h daily for 5 consecutive days. Values are mean +/− SEM; n = 4. *$p > 0.05$ (ANOVA and Dunnett's test). Source: Modified from Tissue Engineering, Mary Ann Liebert (2001)

stimulate osteoblast differentiation. Prior studies using different cell lines and loading protocols yielded contrasting results, likely due to variation in cell type and loading regimen (Klein-Nulend et al. 1997; Kubota et al. 1993). Our findings confirm that specific mechanical conditions can upregulate alkaline phosphatase, potentially enhancing mineral deposition (Bilezikian et al. 2002) in engineered bone tissue.

In conclusion, cyclic pressure (10–40 kPa at 1.0 Hz) applied for 1 h daily over 5 days significantly influences osteoblast proliferation and differentiation markers. These results offer valuable insight into the mechanobiological factors critical to bone tissue engineering and underscore the importance of mechanical stimulus design in bioreactor systems.

2.3 Effect of Cyclic Pressure on Bone Marrow Cell-Derived Osteoclast in Vitro

2.3.1 *Introduction*

As explored in earlier sections (Sects. 2.1 and 2.2), mechanical stress on the skeletal system influences bone balance, which is regulated through a coordinated process involving osteoblast-driven bone formation and osteoclast-mediated bone resorption. Notable examples include reductions in bone density during prolonged immobilization or microgravity environments (e.g., bed rest or space missions) (Zerwekh et al. 1998; Leblanc et al. 1990) and increases in bone mass following regular physical activity (Courteix et al. 1998; Krølner et al. 1983). Despite substantial evidence linking biomechanical loading to skeletal health, the specific cellular and biochemical pathways remain only partially understood.

Initial investigations primarily examined how mechanical forces affect osteoblast behavior in vitro, with results showing that these cells respond to diverse mechanical inputs such as hydrostatic pressure (Ozawa et al. 1990; Roelofsen et al. 1995), shear stress (Reich et al. 1990), and substrate strain (Stanford et al. 1995). However, as bone remodeling involves both osteoblasts and osteoclasts, it is essential to also investigate how these same stimuli impact osteoclast biology. Advances in cell culture techniques have enabled controlled studies of osteoclast precursors under mechanical stimuli (Rubin et al. 1997; Rubin et al. 1999; McAllister et al. 2000; Tsuzuki et al. 2000; Kurata et al. 2001). Prior work demonstrated that osteoclast formation is inhibited by constant hydrostatic pressure (Rubin et al. 1997) or cyclic mechanical strain (Rubin et al. 1999), while mechanical stress also modulates signaling molecules like nitric oxide and prostaglandins in osteoclast-like cells (McAllister et al. 2000) These findings highlight the complex interplay between physical forces and osteoclast function, although the role of such forces in modulating osteoclast progenitor activity within bone marrow remains underexplored.

In this study, we hypothesized that cyclic pressure influences bone resorption by modulating precursor differentiation and cytokine expression. To investigate this, we used a custom apparatus to apply cyclic pressure to rat bone marrow cell cultures and assessed effects on osteoclastogenesis, resorptive function, and cytokine mRNA expression.

2.3.2 *Summary of Experimental Results*

Bone marrow, a rich source of osteoclast progenitors, was extracted from the femurs of Sprague-Dawley rats using previously validated protocols (Hata et al. 1992). Whole marrow cultures were treated with 1,25-dihydroxyvitamin D_3 and either kept under static conditions or exposed to dynamic pressure cycles (10–40 kPa at 1.0 Hz) for 1 h daily for up to 7 days. Control samples remained static throughout. At designated time points, cells were stained for TRAP, an osteoclast marker, and counted. TRAP enzymatic activity was also quantified (Lieberherr et al. 1973). Cultures exposed to cyclic pressure early (days 1–7) displayed significantly fewer TRAP-positive cells than static controls ($p < 0.01$ and $p < 0.05$, Fig. 2.6a), whereas delayed pressure application (starting after 7 days of static culture) had no significant effect on osteoclast numbers (Fig. 2.6b).

In a parallel experiment using devitalized bone slices, TRAP activity served as a proxy for osteoclast population size. Both static and delayed-loading cultures showed similar TRAP levels (Fig. 2.7a). However, surface analysis of resorption lacunae (bone degradation pits) revealed that cyclic pressure significantly ($p < 0.05$) reduced total resorption area (Fig. 2.7b).

Additionally, non-osteoclastic cells were analyzed for cytokine mRNA levels using RNAse protection assays, with results normalized to ribosomal protein L32. Expression of TGF-β1 and IL-6 remained mostly unchanged (Fig. 2.8a–d). However, levels of IL-1α, IL-1β, and TNF-α were consistently reduced under cyclic pressure, suggesting a suppressive effect on pro-osteoclastogenic signaling (Fig. 2.8a–d).

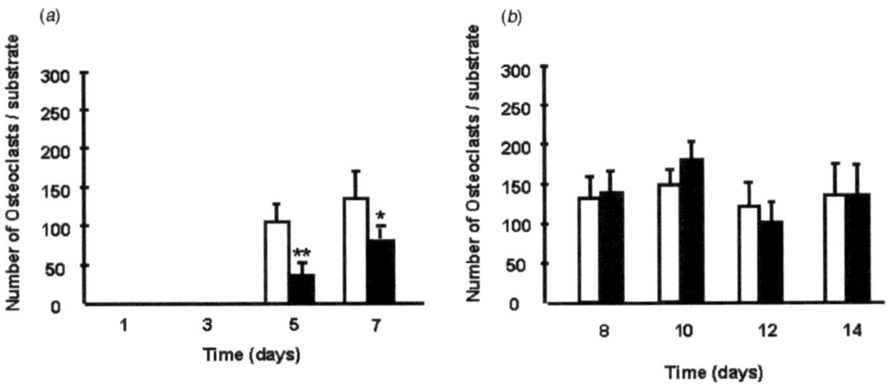

Fig. 2.6 Effect of Timing of Cyclic Pressure on Osteoclast Formation from Bone Marrow Cells. No TRAP-positive osteoclasts were observed in cultures after just 1 or 3 days of incubation. When cyclic pressure (10–40 kPa at 1.0 Hz, 1 h/day) was applied from the start of culture, a statistically significant reduction in TRAP-positive cell numbers was observed after 5 and 7 days compared to static controls (**$p < 0.01$, *$p < 0.05$) (panel a). However, applying cyclic pressure after an initial 7-day static culture period did not significantly affect osteoclast formation, as similar numbers of TRAP-positive cells were found in both control and pressure-treated groups (panel b). (Adapted with permission from *J. Biomech. Eng.*, ASME (2002))

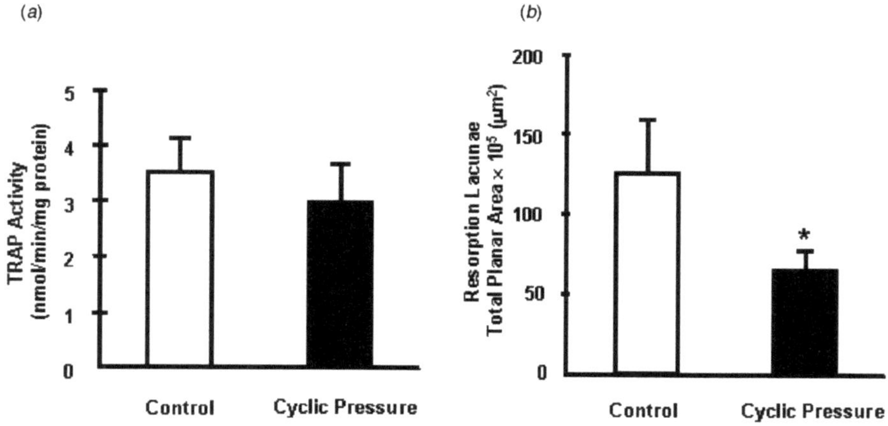

Fig. 2.7 Influence of Cyclic Pressure on Osteoclast Resorption Activity on Bone Slices. (**a**) TRAP activity levels, an indicator of osteoclast presence, remained similar in cultures maintained statically versus those treated with cyclic pressure following a 7-day pre-culture period. (**b**) Total resorbed surface area, assessed by toluidine blue staining of bone slices, was significantly reduced (*$p < 0.05$) in cultures exposed to cyclic pressure (10–40 kPa at 1.0 Hz for 1 h/day over 7 days), suggesting impaired resorptive function despite unchanged osteoclast numbers. (Adapted with permission from *J. Biomech. Eng.*, ASME (2002))

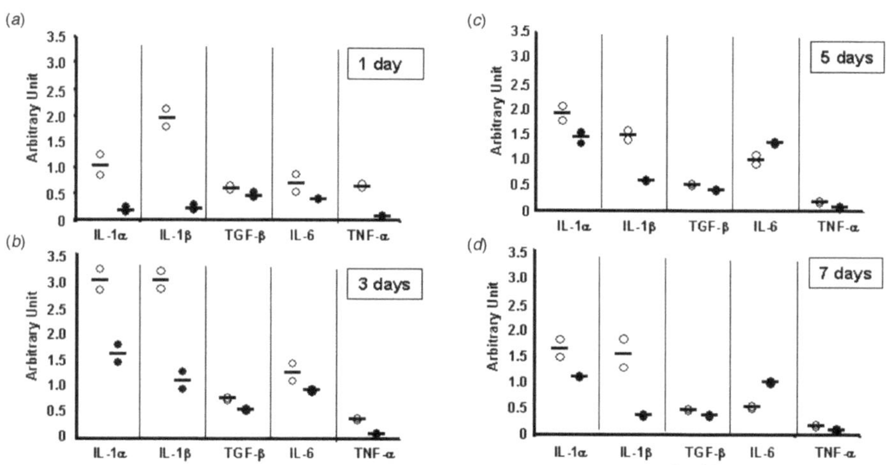

Fig. 2.8 Cytokine mRNA Expression in Bone Marrow Cells Following Cyclic Pressure Exposure. (**a–d**) Expression of TGF-β1 remained unchanged throughout the pressure exposure periods. IL-6 expression declined after 1 and 3 days of loading (**a, b**), but increased with 5 and 7 days of exposure (**c, d**). mRNA levels of IL-1α, IL-1β, and TNF-α were consistently reduced in pressure-exposed groups compared to static controls across all time points, indicating a suppressive effect on pro-inflammatory cytokine signaling. (Adapted with permission from *J. Biomech. Eng.*, ASME (2002))

2.3.3 Discussion

The applied cyclic pressure (10–40 kPa) was intended to mimic physiological conditions within bone marrow cavities. A key observation was that early application of this mechanical stimulus (within 24 h of cell isolation) significantly suppressed osteoclast differentiation (Fig. 2.6a), while delayed exposure did not alter cell numbers (Fig. 2.6b). This temporal dependence supports findings from earlier research showing that continuous (Rubin et al. 1997) or intermittent mechanical (Klein-Nulend et al. 1990) loading reduces osteoclast formation. Since osteoclasts originate from mononuclear precursors that undergo fusion (Suda et al. 1992), it is plausible that cyclic pressure disrupts either the differentiation process or the fusion step. Previous findings reporting an increase in TRAP-positive mononuclear—but not multinuclear—cells under continuous pressure (Imamura et al. 1990) support the latter hypothesis.

Another noteworthy result was the decline in bone resorptive function when cyclic pressure was applied during the second week of culture (Fig. 2.7). Importantly, this occurred without a reduction in osteoclast numbers, implying that cyclic pressure impairs osteoclast activity rather than formation. Given that bone marrow cultures include various non-osteoclastic cells, it is possible that mechanical pressure indirectly affects osteoclasts via altered cytokine secretion. Existing evidence shows that cytokines like IL-1 and TNF-α promote osteoclast differentiation and activity (Manolagas and Jilka 1995; Goldring and Goldring 1996), and blocking these molecules reduces bone resorption (Kimble et al. 1995; Tani-Ishii et al. 1999). Our findings of decreased IL-1α, IL-1β, and TNF-α expression under pressure (Fig. 2.8) accompanied by reduced osteoclastogenesis and resorption (Figs. 2.6 and 2.7) support a regulatory role for these cytokines in mediating mechanical effects.

This study demonstrates that cyclic pressure applied to bone marrow cultures suppresses osteoclast development and diminishes bone resorptive function. These effects are accompanied by reduced expression of key cytokines involved in osteoclast activation. Together with earlier findings on osteoblast response to cyclic pressure, this work contributes to a more complete understanding of how mechanical loading influences skeletal homeostasis at the cellular and molecular levels.

2.4 Cyclic Pressure Affects Collagen Synthesis and Calcium Accumulation by Osteobasts In Vitro

2.4.1 Introduction

In Sect. 2.3, the study focused on the bone resorption aspect of bone homeostasis that is influenced by mechanical stimuli. However, the bone mass balance may shift to the positive side due to increased mechanical loading on the skeleton. For instance, the bones of gymnasts that were routinely subjected to high-impact loading exhibited increased mineral density compared to those of swimmers who performed nonimpact sport (Courteix et al. 1998). In addition, rats subjected to daily exercise regimes of running on a treadmill exhibited greater new bone formation compared to sedentary controls (Iwamoto et al. 1999). Although these studies demonstrated that bone formation in vivo was influenced by external mechanical loading, the cellular/molecular mechanism(s) that link mechanical stimuli on the skeleton and biological events, such as new bone formation by osteoblasts, are not explained. Although a number of in vitro studies examined the effect of various forms of mechanical stimuli such as fluid shear stress (Reich et al. 1990), stretching (Stanford et al. 1995; Vadiakas and Banes 1992), and pressure (Roelofsen et al. 1995) on aspects of osteoblast function, the correlations between mechanical loading and functions of osteoblasts pertinent to new bone formation were not fully understood. New bone formation by osteoblasts is multi-staged and involves synthesis of collagenous (mainly type-I collagen) and non-collagenous proteins (such as osteopontin, osteocalcin, osteonectin, and bone sialoprotein) to form the extracellular matrix. This is followed by incorporation of inorganic minerals, such as calcium phosphate, to form the mineralized bone matrix (Cowin 2001). These processes are regulated by a host of hormones and cytokines, such as parathyroid hormone, transforming growth factor-β, and bone morphogenetic proteins, which are known to affect osteoblast functions (Cowin 2001). Thus, to improve understanding of the relationship between mechanical loading and new bone formation, it is necessary to holistically examine both the osteoblast function (production of the organic phase of bone) and events in their extracellular matrix (the mineralization process). The objective of this study was to test the hypothesis that cyclic pressure stimulates osteoblast function pertinent to new bone formation. Using a custom device, cultures of osteoblasts were exposed to this mechanical stimulus and its effects on the mRNA expression for type-I collagen and osteocalcin by these cells were examined. In addition, collagen and calcium contents in these osteoblast cultures following exposure to cyclic pressure were quantified.

2.4.2 Summary of Experimental Results

Osteoblasts were isolated from the calvariae of neonatal Sprague-Dawley rats and cultured according to published protocols (Nagatomi et al. 2001). Osteoblasts (passage numbers 1–3) in in DMEM containing 10% fetal bovine serum, 10 mM β-glycerophosphate, and 50 mg/mL ascorbic acid were used in the experiments. The phenotype of the osteoblasts was established by the mRNA expression for type-I collagen, alkaline phosphatase, osteopontin, and osteocalcin determined using the reverse transcriptase–polymerase chain reaction (RT-PCR) (Arulanandam et al. 2000). Expression of alkaline phosphatase in situ was determined using a commercially available kit and accumulation of calcium–phosphate minerals in the extracellular matrix of osteoblasts was determined using the von Kossa staining method (Krajian et al. 1952).

Osteoblasts were seeded (17,000 cells/cm^2) on either tissue-culture plastic (polystyrene) plates (60 mm diam) or etched glass coverslips (18 mm diam) and maintained under standard cell culture conditions for 2 days. These cells were then exposed to cyclic pressure in the range 10–40 kPa (above atmospheric pressure) at 1.0 Hz frequency, for 1 h daily for 5, 12, and 19 consecutive days. Controls were prepared in a similar manner from the same batch of osteoblasts but maintained under static conditions for the duration of the experiments. During experiments, the culture medium was changed every 2 days. At the end of the prescribed time periods, various cellular/molecular responses of osteoblasts was examined.

One microgram of total RNA from each sample was reverse transcribed into cDNA using a commercially available kit. The resulting cDNA was amplified using the polymerase chain reaction (PCR) technique following established protocols (Arulanandam et al. 2000) with specific primers for rat osteocalcin and type-I collagen (αI). Hypoxanthine phosphoribosyltransferase (HPRT) was used as a house keeping gene to ensure equal loading of RNA in all PCR reactions. Sense and antisense primers for PCR were designed using pertinent mRNA sequences for these genes as listed in the NIH GenBank database and had the following sequences: osteocalcin, 5′-CCAAGCAGGAGGGCAGTAA-3′, and 5-′GGGCTCCAGG GCAACACAT-3′; type-I collagen (αI) 5′-TCGTGTAAACTCCCTCCATCC-3′, and 5′-AAGGAACAGAAAAGGCAGCAT-3′; HPRT, 5′-TCTTTGCTGACCTGCT GGATT-3′ and 5′-GCTTTTCCACTTTCGCTGATG-3′. PCR amplification was performed by thermal cycling of samples at 95 °C for 5 min and then subjected to the following amplification profile: 1 min at 95 °C, 1 min at 65 °C, and 1 min at 72 °C for a duration of 25 cycles. This series of cycles was followed by a final extension at 72 °C for 10 min. Subsequently, the PCR products were separated on a 2.5% agarose gel, stained with ethidium bromide and visualized by UV transillumination.

Acid-soluble collagen was quantified using a commercially available collagen kit and methods adapted from the literature (Cunnigham Lwaf 1982). Specifically, aliquots (50 μL) of solubilized collagen obtained from each experimental specimen were mixed with 1 mL of dye reagent and the precipitated collagen-dye complex was sedimented by centrifugation. Each resulting pellet was then solubilized in 1 mL of 1.0 N NaOH and light absorbance of the alkali-dye solutions was measured at a wavelength of 570 nm. Collagen concentration of each sample was determined using standard curves prepared with known concentrations of type-I collagen in parallel with the cell lysate samples.

Calcium concentration in the extracellular matrix of osteoblast cultures was quantified using methods adapted from the literature (Jaiswal et al. 1997). Briefly, cells were lysed with 0.5 N HCl at 4 °C overnight and, following centrifugation, the clear supernatant was analyzed using a commercially available calcium assay kit. Light absorbance of each sample–reagent mixture was measured at a wavelength of 570 nm. Calcium concentration of each sample was determined using standard curves prepared with known concentrations of calcium in parallel with experimental samples.

Numerical data were obtained from experiments that were run in duplicates and repeated (at least) three separate times. The level of significance in differences between control and experimental groups was determined by performing Student t-tests; p values less than 0.05 were considered statistically significant.

RT-PCR amplification of mRNA revealed that osteocalcin mRNA expression was similar in osteoblasts that were either maintained under control conditions or exposed to the cyclic pressure tested for 5, 12, and 19 consecutive days (Fig. 2.9). Type-I collagen mRNA expression was similar in controls and in osteoblast cultures exposed to this cyclic pressure for 1 h daily for 5 and 12 days. In contrast, and compared to controls, type-I collagen mRNA expression was increased when these cell cultures were exposed to cyclic pressure for 19 consecutive days (Fig. 2.9).

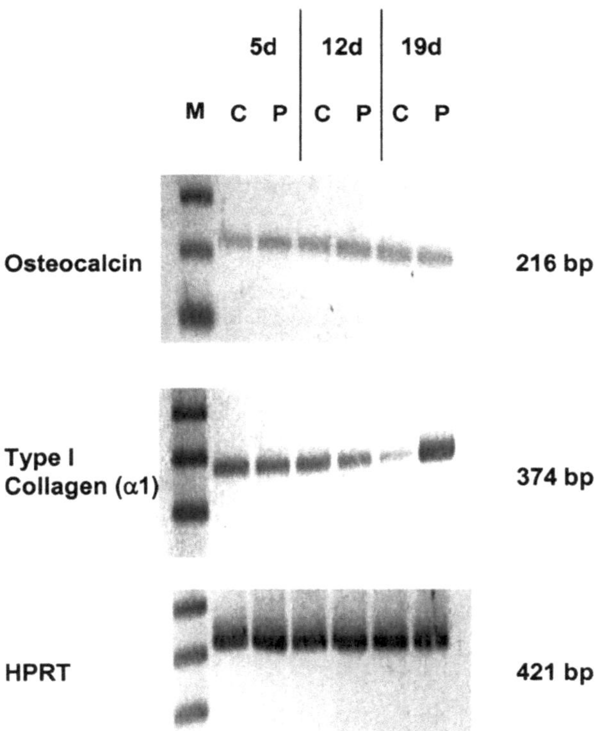

Fig. 2.9 Time-course mRNA expression for select bone proteins by osteoblasts exposed to cyclic pressure. Expression of mRNA for the noncollagenous protein osteocalcin, was similar in controls and in cells exposed to cyclic pressure after 5, 12, and 19 days. Type-I collagen mRNA expression was similar in controls and osteoblasts exposed to cyclic pressure for 5 and 12 days; in contrast, and compared to controls, type-I collagen mRNA expression increased when osteoblast cultures were exposed to cyclic pressure for 19 consecutive days. C: osteoblasts maintained under control static conditions. P: osteoblasts exposed to cyclic pressure (10–40 kPa at 1.0 Hz. for 1 h daily) M: base pair marker. HPRT: hypoxanthine phosphoribosyl transferase (housekeeping gene)

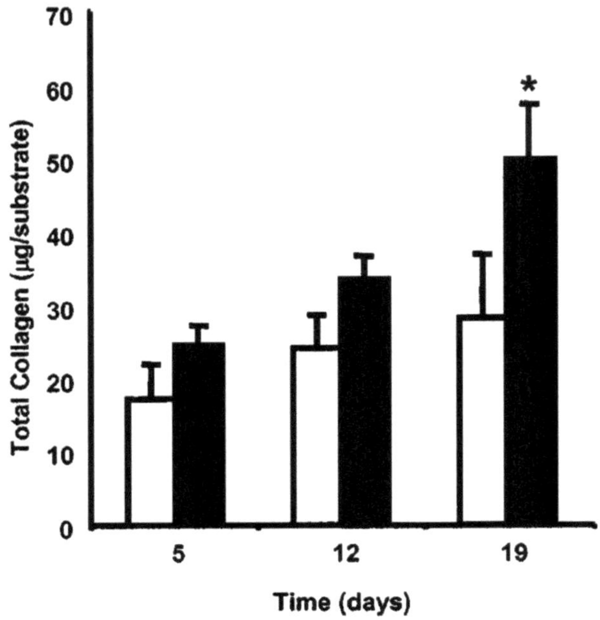

Fig. 2.10 Collagen content in osteoblast cultures exposed to cyclic pressure. After 5 and 12 days, collagen concentrations were similar in controls (white) and in osteoblasts exposed to cyclic pressure (black) for 1 h daily. After 19 days, however, compared to controls, collagen concentration was significantly ($p < 0.05$) higher in osteoblast cultures exposed to cyclic pressure. Values are mean +/− S.E.M.; n = 3; *p < 0.05 (t-test)

After 5 and 12 days, the amount of acid-soluble collagen was similar in controls and in osteoblast cultures exposed to cyclic pressure (10–40 kPa at 1 Hz frequency) for 1 h daily. In contrast, and compared to controls, the amount of acid-soluble collagen in osteoblasts exposed to cyclic pressure was significantly ($p < 0.05$) higher after 19 days (Fig. 2.10).

After 5 and 12 days, calcium content was similar in controls and in osteoblast cultures exposed to cyclic pressure (10–40 kPa at 1 Hz frequency) for 1 h daily (Fig. 2.11). After 19 days, however, and compared to controls, the calcium content was significantly ($p < 0.05$) higher in osteoblast cultures exposed to cyclic pressure (Fig. 2.11).

2.4.3 Discussion

This study aimed to simulate the mechanical loading conditions to which osteoblasts are subjected in vivo. The cyclic pressure parameters (10 kPa minimum: 40 kPa maximum at 1.0 Hz) used in the experiments were based on estimates from macroscopic-level physiological values reported in the literature. For example,

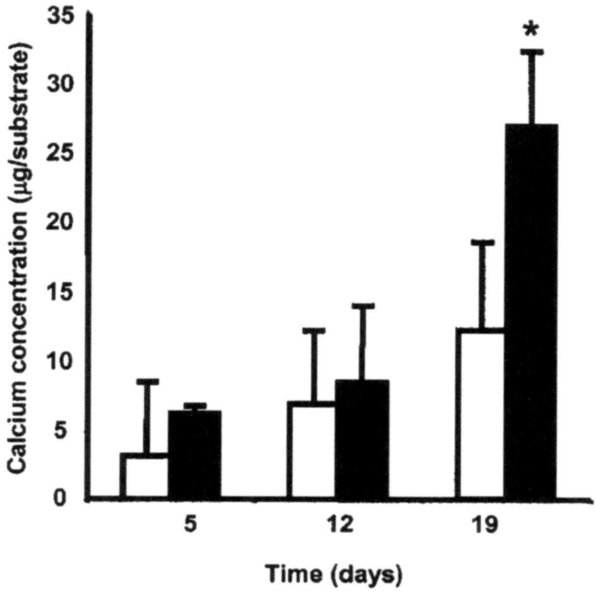

Fig. 2.11 Calcium concentration in osteoblast cultures exposed to cyclic pressure. After 5 and 12 days, the concentrations of calcium in controls (white) and in osteoblast cultures exposed to cyclic pressure (black) for 1 h daily were similar. After 19 days, however, compared to controls, calcium concentrations were significantly ($p < 0.05$) increased when osteoblast cultures were exposed to the cyclic pressure regime tested. Values are mean +/− S.E.M.; $n = 3$; *$p < 0.05$ (t-test)

during walking the hip joint of a 70 kg human experiences approximately 3800 N (Röhrle et al. 1984), which, in turn, can generate up to 210 mmHg (28 kPa) fluid pressure within the bone marrow (Downey et al. 1988). Under the repetitive loading cycles during walking, therefore, osteoblasts inside the femoral head (on the surfaces of trabecular bone adjacent to the bone marrow in vivo) are constantly exposed to cyclic pressures in the order of 10–40 kPa.

Exposure of osteoblast cultures to cyclic pressure resulted in mRNA expression (Fig. 2.9) and synthesis/deposition of collagen (Fig. 2.10) that were sustained for all time points tested. As both of these were attenuated under control (static) conditions after 19 days (Figs. 2.10 and 2.11, respectively), it may be speculated that the critical effect of this mechanical stimulus is to maintain osteoblasts functionally active and productive. Although differences in experimental conditions do not allow direct comparison, these results are in agreement with other reports in the literature that, over an 8-day period, type-I collagen mRNA expression of mouse calvaria osteoblasts decreased under control (no pressure) conditions but increased when these cells were exposed to intermittent hydrostatic compression (13 kPa at 0.3 Hz) (Roelofsen et al. 1995). The new evidence from this study was that exposure to cyclic pressure stimulated not only mRNA expression, but also synthesis/release of collagen by osteoblasts. The increased amount of collagen under cyclic pressure measured in this study (Fig. 2.10), however, could have also resulted from an increased number of osteoblasts instead of increased synthesis of this protein. Since the amount of total DNA was similar for controls and for osteoblast cultures exposed to cyclic pressure at all time points tested (data not presented), and the housekeeping gene expression ensured equal loading of samples in RT-PCR reactions, it was concluded that exposure of osteoblasts to cyclic pressure indeed led to increased synthesis of collagen (the main constituent of the organic phase of bone). These results, therefore, imply that increased collagen synthesis induced by cyclic pressure may be part of the mechanism(s) involved in enhancing bone formation under skeletal loading in vivo.

In addition to extracellular matrix proteins, bone in vivo is composed of calcium-containing mineral. Accumulation of calcium into the extracellular matrix of osteoblasts is, therefore, one of the critical events in new bone formation. One of the main contributions of this study was to quantify the enhanced calcium accumulation in osteoblast cultures exposed to cyclic pressure (Fig. 2.11). These cellular/molecular level findings shed some light on previous reports from a tissue level study that exposure of embryonic (17 day old) mouse calvaria to intermittent hydrostatic pressure (13 kPa at 0.3 Hz) for five consecutive days resulted in enhanced calcification (Klein-Nulend et al. 1987). Although the exact mechanisms that regulate accumulation of minerals in the extracellular matrix of bone are not fully understood, collagen is considered one of the regulators of matrix mineralization in bone (Cowin 2001). For this reason, upregulation of type-I collagen mRNA (Fig. 2.9) as well as the increased collagen (Fig. 2.10) and calcium (Fig. 2.11) content under cyclic pressure observed in this in vitro study support the view that an increase in collagen synthesis may provide a scaffold for increased mineralization of the extracellular matrix.

In addition to collagen, other chemical constituents of the organic of bone (e.g. noncollagenous matrix proteins such as osteocalcin, osteopontin, and bone sialoprotein) are involved in regulating the bone matrix mineralization process (Cowin 2001). More specifically, osteocalcin is considered an inhibitor of matrix mineralization. In this study, osteocalcin mRNA expression was not affected by the cyclic pressure stimulus at all time points tested (Fig. 2.9). Thus, it can be concluded that suppression of inhibitors of matrix mineralization may not be the controlling mechanism of the increased calcium concentration in osteoblast cultures exposed to cyclic pressure. Further research, however, is needed to elucidate the exact relationship(s) between mechanical loading and bone matrix mineralization.

In summary, the results of this in vitro study demonstrated that short duration (1 h) but daily exposure to cyclic pressure for 19 consecutive days enhanced osteoblast functions pertinent to new bone formation by stimulating both synthesis and deposition of collagen, a process which was accompanied by increased accumulation of calcium. This evidence provides molecular-level insight into the mechanism(s) involved in new bone formation under mechanical loading.

References

Acevedo AD, Bowser SS, Gerritsen ME, Bizios R (1993) Morphological and proliferative responses of endothelial cells to hydrostatic pressure: role of fibroblast growth factor. J Cell Physiol 157(3):603–614. https://api.istex.fr/ark:/67375/WNG-477MB61D-8/fulltext.pdf. https://doi.org/10.1002/jcp.1041570321

Arulanandam BP, Mittler JN, Lee WT, O'Toole M, Metzger DW (2000) Neonatal administration of IL-12 enhances the protective efficacy of antiviral vaccines. J Immunol (1950) 164(7):3698–3704. http://www.jimmunol.org/cgi/content/abstract/164/7/3698. https://doi.org/10.4049/jimmunol.164.7.3698

Atkins GB, Jain MK (2007) Role of krüppel-like transcription factors in endothelial biology. Circ Res 100(12):1686–1695. https://doi.org/10.1161/01.RES.0000267856.00713.0a

Aw Yong KM, Horst E, Neale D et al (2022) A bioreactor for 3D in vitro modeling of the mechanical stimulation of osteocytes. Front Bioeng Biotechnol 10:797542. https://www.ncbi.nlm.nih.gov/pubmed/35402411. https://doi.org/10.3389/fbioe.2022.797542

Bancroft GN, Sikavitsas VI, van den Dolder J et al (2002) Fluid flow increases mineralized matrix deposition in 3D perfusion culture of marrow stromal osteoblasts in a dose-dependent manner. Proc Natl Acad Sci 99(20):12600–12605. https://www.jstor.org/stable/3073269. https://doi.org/10.1073/pnas.202296599

Bateman TA, Dunstan CR, Ferguson VL, Lacey DL, Ayers RA, Simske SJ (2000) Osteoprotegerin mitigates tail suspension-induced osteopenia. Bone (New York, N.Y.) 26(5):443–449. https://doi.org/10.1016/S8756-3282(00)00256-8

Bergula AP, Huang W, Frangos JA (1999) Femoral vein ligation increases bone mass in the hindlimb suspended rat. Bone (New York, N.Y.) 24(3):171–177. https://doi.org/10.1016/S8756-3282(98)00165-3

Beverly M, Marks BE, Murray DW (2020) Subchondral pressures and perfusion during weight bearing. J Orthop Surg Res 15(1):239. https://www.ncbi.nlm.nih.gov/pubmed/32600340. https://doi.org/10.1186/s13018-020-01754-y

Bilezikian JP, Raisz LG, Rodan GA (2002) Principles of bone biology. Elsevier Science & Technology, San Diego. https://ebookcentral.proquest.com/lib/[SITE_ID]/detail.action?docID=305643

Brighton C, Fisher JR, Levine S et al (1996) The biochemical pathway mediating the proliferative response of bone cells to a mechanical stimulus. J Bone Joint Surg Am 78(9):1337–1347. http://ovidsp.ovid.com/ovidweb.cgi?T=JS&NEWS=n&CSC=Y&PAGE=fulltext&D=ovft&AN=00004623-199609000-00007. https://doi.org/10.2106/00004623-199609000-00007

Cheng B, Zhao S, Luo J, Sprague E, Bonewald LF, Jiang JX (2001) Expression of functional gap junctions and regulation by fluid flow in osteocyte-like MLO-Y4 cells. J Bone Miner Res 16(2):249–259. https://doi.org/10.1359/jbmr.2001.16.2.249

Chistiakov DA, Orekhov AN, Bobryshev YV (2017) Effects of shear stress on endothelial cells: go with the flow. Acta Physiol 219(2):382–408. https://doi.org/10.1111/apha.12725

Chiu J, Usami S, Chien S (2009) Vascular endothelial responses to altered shear stress: pathologic implications for atherosclerosis. Annals Med (Helsinki) 41(1):19–28. https://doi.org/10.1080/07853890802186921

Conway D, Schwartz MA (2012) Lessons from the endothelial junctional mechanosensory complex. F1000 Biol Rep 4:1. https://www.ncbi.nlm.nih.gov/pubmed/22238515. https://doi.org/10.3410/B4-1

Courteix D, Lespessailles E, Loiseau Peres S, Obert P, Germain P, Benhamou CL (1998) Effect of physical training on bone mineral density in prepubertal girls: a comparative study between impact-loading and non-impact-loading sports. Osteoporos Int 8(2):152–158. https://www.ncbi.nlm.nih.gov/pubmed/9666939. https://doi.org/10.1007/BF02672512

Cowin SC (2001. http://bvbr.bib-bvb.de:8991/F?func=service&doc_library=BVB01&local_base=BVB01&doc_number=009880981&sequence=000002&line_number=0001&func_

code=DB_RECORDS&service_type=MEDIA) Bone mechanics handbook, 2nd edn. CRC Press, Boca Raton. https://doi.org/10.1201/b14263
Cunnigham Lwaf DW (ed) (1982) Methods in enzymology. Academic, New York
Dekker RJ, van Soest S, Fontijn RD et al (2002) Prolonged fluid shear stress induces a distinct set of endothelial cell genes, most specifically lung krüppel-like factor (KLF2). Blood 100(5):1689–1698. https://doi.org/10.1182/blood-2002-01-0046
Dewey CF, Bussolari SR, Gimbrone MA, Davies PF (1981) The dynamic response of vascular endothelial cells to fluid shear stress. J Biomech Eng 103(3):177–185. https://doi.org/10.1115/1.3138276
Downey D, Simkin P, Taggart R (1988) The effect of compressive loading on intraosseous pressure in the femoral head in vitro. J Bone Joint Surgery Am 70(6):871–877. http://ovidsp.ovid.com/ovidweb.cgi?T=JS&NEWS=n&CSC=Y&PAGE=fulltext&D=ovft&AN=00004623-198870060-00012. https://doi.org/10.2106/00004623-198870060-00012
Friedrich EE, Hong Z, Xiong S et al (2019) Endothelial cell Piezo1 mediates pressure-induced lung vascular hyperpermeability via disruption of adherens junctions. Proc Natl Acad Sci 116(26):12980–12985. https://www.jstor.org/stable/26744131. https://doi.org/10.1073/pnas.1902165116
Gandhi J, Kao S, Roux B et al (2019) Perfusion bioreactor culture of bone marrow stromal cells enhances cranial defect regeneration. Plast Reconstr Surg (1963) 143(5):993e–1002e. http://ovidsp.ovid.com/ovidweb.cgi?T=JS&NEWS=n&CSC=Y&PAGE=fulltext&D=ovft&AN=00006534-201905000-00023. https://doi.org/10.1097/PRS.0000000000005529
Gardinier JD, Majumdar S, Duncan RL, Wang L (2009) Cyclic hydraulic pressure and fluid flow differentially modulate cytoskeleton re-organization in MC3T3 osteoblasts. Cell Mol Bioeng 2(1):133–143. https://doi.org/10.1007/s12195-008-0038-2
Genetos DC, Kephart CJ, Zhang Y, Yellowley CE, Donahue HJ (2007) Oscillating fluid flow activation of gap junction hemichannels induces atp release from MLO-Y4 osteocytes. J Cell Physiol 212(1):207–214. https://api.istex.fr/ark:/67375/WNG-DK2LZDDB-7/fulltext.pdf. https://doi.org/10.1002/jcp.21021
Goldring SR, Goldring MB (1996) Cytokines and skeletal physiology. Clinical Orthopaedics and Related Research 324(324):13–23
Guo D, Liang S, Wang S et al (2016) Role of epithelial na+ channels in endothelial function. J Cell Sci 129(2):290–297. https://www.ncbi.nlm.nih.gov/pubmed/26621031. https://doi.org/10.1242/jcs.168831
Gurkan UA, Akkus O (2008) The mechanical environment of bone marrow: a review. Ann Biomed Eng 36(12):1978–1991. https://doi.org/10.1007/s10439-008-9577-x
Hall JE, Guyton AC (2016) Guyton and hall textbook of medical physiology, 13th edn. Elsevier, Philadelphia, PA. http://bvbr.bib-bvb.de:8991/F?func=service&doc_library=BVB01&local_base=BVB01&doc_number=028186518&sequence=000003&line_number=0001&func_code=DB_RECORDS&service_type=MEDIA
Han Y, Cowin SC, Schaffler MB, Weinbaum S (2004) Mechanotransduction and strain amplification in osteocyte cell processes. Proc Natl Acad Sci 101(47):16689–16694. https://www.jstor.org/stable/3373886. https://doi.org/10.1073/pnas.0407429101
Hastings NE, Simmers MB, McDonald OG, Wamhoff BR, Blackman BR (2007) Atherosclerosis-prone hemodynamics differentially regulates endothelial and smooth muscle cell phenotypes and promotes pro-inflammatory priming. Am J Physiol Cell Physiol 293(6):1824–1833. http://ajpcell.physiology.org/content/293/6/C1824. https://doi.org/10.1152/ajpcell.00385.2007
Hata K, Kukita T, Akamine A, Kukita A, Kurisu K (1992) Trypsinized osteoclast-like multinucleated cells formed in rat bone marrow cultures efficiently form resorption lacunae on dentine. Bone (New York, N.Y.) 13(2):139–146. https://doi.org/10.1016/8756-3282(92)90003-F
Hergenreider E, Heydt S, Treguer K et al (2012) Atheroprotective communication between endothelial cells and smooth muscle cells through miRNAs. Nat Cell Biol 14(3):249–256. https://www.narcis.nl/publication/RecordID/oai:pure.atira.dk:publications%2F25579b5f-11ce-4dd0-9313-800039a87e42. https://doi.org/10.1038/ncb2441

Hung CT, Pollack SR, Reilly TM, Brighton CT (1995) Real-time calcium response of cultured bone cells to fluid flow. Clin Orthop Relat Res 313(313):256–269. https://www.ncbi.nlm.nih.gov/pubmed/7641488

Imamura K, Ozawa H, Hiraide T et al (1990) Continuously applied compressive pressure induces bone resorption by a mechanism involving prostaglandin E2 synthesis. J Cell Physiol 144(2):222–228. https://api.istex.fr/ark:/67375/WNG-47WSK9Q3-Q/fulltext.pdf. https://doi.org/10.1002/jcp.1041440207

Ishaug SL, Crane GM, Miller MJ, Yasko AW, Yaszemski MJ, Mikos AG (1997) Bone formation by three-dimensional stromal osteoblast culture in biodegradable polymer scaffolds. J Biomed Mater Res 36(1):17–28

Iwamoto J, Yeh JK, Aloia JF (1999) Differential effect of treadmill exercise on three cancellous bone sites in the young growing rat. Bone (New York, N.Y.) 24(3):163–169. https://doi.org/10.1016/S8756-3282(98)00189-6

Jacobs CR, Yellowley CE, Davis BR, Zhou Z, Cimbala JM, Donahue HJ (1998) Differential effect of steady versus oscillating flow on bone cells. J Biomech 31(11):969–976. https://doi.org/10.1016/S0021-9290(98)00114-6

Jaiswal N, Haynesworth SE, Caplan AI, Bruder SP (1997) Osteogenic differentiation of purified, culture-expanded human mesenchymal stem cells in vitro. J Cell Biochem 64(2):295–312. https://api.istex.fr/ark:/67375/WNG-CVG1SH9Z-D/fulltext.pdf. https://doi.org/10.1002/(SICI)1097-4644(199702)64:23.0.CO;2-I

Johnson DL, McAllister TN, Frangos JA (1996) Fluid flow stimulates rapid and continuous release of nitric oxide in osteoblasts. Am J Physiol Endocrinol Metab 271(1):205–208. http://ajpendo.physiology.org/content/271/1/E205. https://doi.org/10.1152/ajpendo.1996.271.1.E205

Kimble RB, Matayoshi AB, Vannice JL, Kung VT, Williams C, Pacifici R (1995) Simultaneous block of interleukin-1 and tumor necrosis factor is required to completely prevent bone loss in the early postovariectomy period. Endocrinology (Philadelphia) 136(7):3054–3061. https://www.ncbi.nlm.nih.gov/pubmed/7789332. https://doi.org/10.1210/en.136.7.3054

Kitamura, Hasegawa, Keigo et al (2017) Piezo type mechanosensitive ion channel component 1 functions as a regulator of the cell fate determination of mesenchymal stem cells. Springer

Klein-Nulend J, Veidhuijzen JP, De Jong M, Burger EH (1987) Increased bone formation and decreased bone resorption in fetal mouse calvaria as a result of intermittent compressive force in vitro. Bone Miner 2(6):441–448. https://www.ncbi.nlm.nih.gov/pubmed/3505768

Klein-Nulend J, Paul Veldhuijzen J, Van Strien ME, Jong MD, Burger EH (1990) Inhibition of osteoclastic bone resorption by mechanical stimulation in vitro. Arthritis Rheum 33(1):66–72. https://doi.org/10.1002/art.1780330108

Klein-Nulend J, Roelofsen J, Sterck JGH, Semeins CM, Burger EH (1995) Mechanical loading stimulates the release of transforming growth factor-β activity by cultured mouse calvariae and periosteal cells. J Cell Physiol 163(1):115–119. https://api.istex.fr/ark:/67375/WNG-5H4FQZ8D-5/fulltext.pdf. https://doi.org/10.1002/jcp.1041630113

Klein-Nulend J, Roelofsen J, Semeins CM, Bronckers ALJJ, Burger EH (1997) Mechanical stimulation of osteopontin mRNA expression and synthesis in bone cell cultures. J Cell Physiol 170(2):174–181. https://api.istex.fr/ark:/67375/WNG-4VP7C1NB-G/fulltext.pdf. https://doi.org/10.1002/(SICI)1097-4652(199702)170:23.0.CO;2-L

KnoTate ML, Steck R, Forwood MR, Niederer P (2000) In vivo demonstration of load-induced fluid flow in the rat tibia and its potential implications for processes associated with functional adaptation. J Exp Biol 203(18):2737–2745. http://jeb.biologists.org/cgi/content/abstract/203/18/2737. https://doi.org/10.1242/jeb.203.18.2737

Kou S, Pan L, van Noort D et al (2011) A multishear microfluidic device for quantitative analysis of calcium dynamics in osteoblasts. Biochem Biophys Res Commun 408(2):350–355. https://doi.org/10.1016/j.bbrc.2011.04.044

Krajian AA, Gradwohl RBH, (Rutherford Birchard Hayes) (1952) Histopathological technic

Krølner B, Toft B, Pors Nielsen S, Tøndevold E (1983) Physical exercise as prophylaxis against involutional vertebral bone loss: a controlled trial. Clin Sci (1979) 64(5):541–546. https://www.ncbi.nlm.nih.gov/pubmed/6339154. https://doi.org/10.1042/cs0640541

Kubota T, Yamauchi M, Onozaki J, Sato S, Suzuki Y, Sodek J (1993) Influence of an intermittent compressive force on matrix protein expression by ROS 17/2.8 cells, with selective stimulation of osteopontin. Arch Oral Biol 38(1):23–30. https://www.ncbi.nlm.nih.gov/pubmed/8382921. https://doi.org/10.1016/0003-9969(93)90150-K

Kuebler WM, Ying X, Bhattacharya J (2002) Pressure-induced endothelial ca(2+) oscillations in lung capillaries. Am J Physiol Lung Cell Mol Physiol 282(5):L917–L923. https://www.ncbi.nlm.nih.gov/pubmed/11943655

Kurata K, Uemura T, Nemoto A et al (2001) Mechanical strain effect on bone-resorbing activity and messenger RNA expressions of marker enzymes in isolated osteoclast culture. J Bone Miner Res 16(4):722–730. https://doi.org/10.1359/jbmr.2001.16.4.722

Lanyon LE, Hampson WGJ, Goodship AE, Shah JS (1975) Bone deformation recorded in vivo from strain gauges attached to the human tibial shaft. Acta Orthop 46(2):256–268. https://doi.org/10.3109/17453677508989216

Leblanc AD, Schneider VS, Evans HJ, Engelbretson DA, Krebs JM, LaBlanc AD (1990) Bone mineral loss and recovery after 17 weeks of bed rest. J Bone Miner Res 5(8):843–850. https://ntrs.nasa.gov/citations/20050000871. https://doi.org/10.1002/jbmr.5650050807

Levesque MJ, Nerem RM (1985) The elongation and orientation of cultured endothelial cells in response to shear stress. J Biomech Eng 107(4):341–347. https://doi.org/10.1115/1.3138567

Li J, Rose E, Frances D, Sun Y, You L (2011) Effect of oscillating fluid flow stimulation on osteocyte mRNA expression. J Biomech 45(2):247–251. https://www.clinicalkey.es/playcontent/1-s2.0-S0021929011006798. https://doi.org/10.1016/j.jbiomech.2011.10.037

Li X, Liu C, Li P et al (2013) Connexin 43 is a potential regulator in fluid shear stress-induced signal transduction in osteocytes. J Orthop Res 31(12):1959–1965. https://api.istex.fr/ark:/67375/WNG-BQB8S2VX-0/fulltext.pdf. https://doi.org/10.1002/jor.22448

Lieberherr M, Vreven J, Vaes G (1973) The acid and alkaline phosphatases, inorganic pyrophosphatases and phosphoprotein phosphatase of bone. I. characterization and assay. Biochim Biophys Acta 293(1):160–169. https://www.ncbi.nlm.nih.gov/pubmed/4346571

Lin Z, Kumar A, SenBanerjee S et al (2005) Kruppel-like factor 2 (KLF2) regulates endothelial thrombotic function. Circ Res 96(5):e48–e57. http://circres.ahajournals.org/cgi/content/abstract/96/5/e48. https://doi.org/10.1161/01.RES.0000159707.05637.a1

Liu C, Zhao Y, Cheung W, Gandhi R, Wang L, You L (2010) Effects of cyclic hydraulic pressure on osteocytes. Bone (New York, N.Y.) 46(5):1449–1456. https://www.clinicalkey.es/playcontent/1-s2.0-S875632821000428X. https://doi.org/10.1016/j.bone.2010.02.006

Man J, Graham T, Squires-Donelly G, Laslett AL (2022) The effects of microgravity on bone structure and function. NPJ Microgravity 8(1):9. https://www.ncbi.nlm.nih.gov/pubmed/35383182. https://doi.org/10.1038/s41526-022-00194-8

Manolagas SC, Jilka RL (1995) Bone marrow, cytokines, and bone remodeling—emerging insights into the pathophysiology of osteoporosis. N Engl J Med 332(5):305–311. http://content.nejm.org/cgi/content/extract/332/5/305. https://doi.org/10.1056/NEJM199502023320506

McAllister TN, Du T, Frangos JA (2000) Fluid shear stress stimulates prostaglandin and nitric oxide release in bone marrow-derived preosteoclast-like cells. Biochem Biophys Res Commun 270(2):643–648. https://doi.org/10.1006/bbrc.2000.2467

McCoy RJ, Widaa A, Watters KM et al (2013) Orchestrating osteogenic differentiation of mesenchymal stem cells—identification of placental growth factor as a mechanosensitive gene with a pro-osteogenic role. Stem Cells 31(11):2420–2431. https://doi.org/10.1002/stem.1482

McGarry JP, Klein Nulend J, Prendergast PJ (2005) The effect of cytoskeletal disruption on pulsatile fluid flow-induced nitric oxide and prostaglandin E2 release in osteocytes and osteoblasts. Biochem Biophys Res Commun 330(1):341–348. https://www.narcis.nl/publication/RecordID/oai:dare.uva.nl:publications%2F8696cf87-b7c1-40e4-82cb-c455161c1a80. https://doi.org/10.1016/j.bbrc.2005.02.175

McKinley M, O'Loughlin V (2005) Human anatomy, 5th edn. McGrow-Hill

Nagatomi J, Arulanandam BP, Metzger DW, Meunier A, Bizios R (2001) Frequency- and duration-dependent effects of cyclic pressure on select bone cell functions. Tissue Eng 7(6):717–728. https://www.ncbi.nlm.nih.gov/pubmed/11749729. https://doi.org/10.1089/107632701753337672

Nilius B, Droogmans G (2001) Ion channels and their functional role in vascular endothelium. Physiol Rev 81(4):1415–1459. http://physrev.physiology.org/content/81/4/1415. https://doi.org/10.1152/physrev.2001.81.4.1415

Ohashi T, Sugaya Y, Sakamoto N, Sato M (2007) Hydrostatic pressure influences morphology and expression of VE-cadherin of vascular endothelial cells. J Biomech 40(11):2399–2405. https://www.clinicalkey.es/playcontent/1-s2.0-S0021929006004817. https://doi.org/10.1016/j.jbiomech.2006.11.023

Ozawa H, Imamura K, Abe E et al (1990) Effect of a continuously applied compressive pressure on mouse osteoblast-like cells (MC3T3-E1) in vitro. J Cell Physiol 142(1):177–185. https://api.istex.fr/ark:/67375/WNG-0L06CN5L-4/fulltext.pdf. https://doi.org/10.1002/jcp.1041420122

Papaioannou TG, Karatzis EN, Vavuranakis M, Lekakis JP, Stefanadis C (2006) Assessment of vascular wall shear stress and implications for atherosclerotic disease. Int J Cardiol 113(1):12–18. https://doi.org/10.1016/j.ijcard.2006.03.035

Pavalko FM, Chen NX, Turner CH et al (1998) Fluid shear-induced mechanical signaling in MC3T3-E1 osteoblasts requires cytoskeleton-integrin interactions. Am J Physiol Cell Physiol 275(6):1591–1601. http://ajpcell.physiology.org/content/275/6/C1591. https://doi.org/10.1152/ajpcell.1998.275.6.c1591

Price C, Zhou X, Li W, Wang L (2011) Real-time measurement of solute transport within the lacunar-canalicular system of mechanically loaded bone: direct evidence for load-induced fluid flow. J Bone Miner Res 26(2):277–285. https://doi.org/10.1002/jbmr.211

Prystopiuk V, Fels B, Simon CS et al (2018) A two-phase response of endothelial cells to hydrostatic pressure. J Cell Sci 131(12):jcs206920. https://www.ncbi.nlm.nih.gov/pubmed/29848657. https://doi.org/10.1242/jcs.206920

Puleo DA, Holleran LA, Doremus RH, Bizios R (1991) Osteoblast responses to orthopedic implant materials in vitro. J Biomed Mater Res 25(6):711–723. https://api.istex.fr/ark:/67375/WNG-G4VWC4FT-V/fulltext.pdf. https://doi.org/10.1002/jbm.820250603

Qi Y, Jiang J, Jiang X et al (2011) PDGF-BB and TGF-β1 on cross-talk between endothelial and smooth muscle cells in vascular remodeling induced by low shear stress. Proc Natl Acad Sci PNAS 108(5):1908–1913. https://agris.fao.org/agris-search/search.do?recordID=US201301937926. https://doi.org/10.1073/pnas.1019219108

Reich KM, Gay CV, Frangos JA (1990) Fluid shear stress as a mediator of osteoblast cyclic adenosine monophosphate production. J Cell Physiol 143(1):100–104. https://api.istex.fr/ark:/67375/WNG-FXWHXWRB-7/fulltext.pdf. https://doi.org/10.1002/jcp.1041430113

Reinhart-King CA, Fujiwara K, Berk BC (2008) Physiologic stress-mediated signaling in the endothelium. Methods Enzymol 443:25–44. https://www.ncbi.nlm.nih.gov/pubmed/18772009. https://doi.org/10.1016/S0076-6879(08)02002-8

Robling AG, Hinant FM, Burr DB, Turner CH (2002) Improved bone structure and strength after long-term mechanical loading is greatest if loading is separated into short bouts. J Bone Miner Res 17(8):1545–1554. https://doi.org/10.1359/jbmr.2002.17.8.1545

Roelofsen J, Klein-Nulend J, Burger EH (1995) Mechanical stimulation by intermittent hydrostatic compression promotes bone-specific gene expression in vitro. J Biomech 28(12):1493–1503. https://doi.org/10.1016/0021-9290(95)00097-6

Röhrle H, Scholten R, Sigolotto C, Sollbach W, Kellner H (1984) Joint forces in the human pelvis-leg skeleton during walking. J Biomech 17(6):409–424. https://doi.org/10.1016/0021-9290(84)90033-2

Rubin J, Biskobing D, Fan X, Rubin C, McLeod K, Taylor WR (1997) Pressure regulates osteoclast formation and MCSF expression in marrow culture. J Cell Physiol 170(1):81–87. https://

api.istex.fr/ark:/67375/WNG-8F2FS8KN-H/fulltext.pdf. https://doi.org/10.1002/(SICI)1097-4652(199701)170:13.0.CO;2-H

Rubin J, Fan X, Biskobing DM, Taylor WR, Rubin CT (1999) Osteoclastogenesis is repressed by mechanical strain in an in vitro model. J Orthop Res 17(5):639–645. https://api.istex.fr/ark:/67375/WNG-6MKPLRQR-3/fulltext.pdf. https://doi.org/10.1002/jor.1100170504

Sakamoto N, Ohashi T, Sato M (2006) Effect of fluid shear stress on migration of vascular smooth muscle cells in cocultured model. Ann Biomed Eng 34(3):408–415. https://www.ncbi.nlm.nih.gov/pubmed/16482415. https://doi.org/10.1007/s10439-005-9043-y

Sakamoto N, Kiuchi T, Sato M (2011) Development of an Endothelial–Smooth muscle cell coculture model using phenotype-controlled smooth muscle cells. Ann Biomed Eng 39(11):2750–2758. https://doi.org/10.1007/s10439-011-0372-8

Salwen SA, Szarowski DH, Turner JN, Bizios R (1998) Three-dimensional changes of the cytoskeleton of vascular endothelial cells exposed to sustained hydrostatic pressure. Med Biol Eng Comput 36(4):520–527. https://www.ncbi.nlm.nih.gov/pubmed/10198540. https://doi.org/10.1007/BF02523225

Schwartz EA, Bizios R, Medow MS, Gerritsen ME (1999) Exposure of human vascular endothelial cells to sustained hydrostatic pressure stimulates proliferation: involvement of the alphav integrins. Circ Res 84(3):315–322. http://ovidsp.ovid.com/ovidweb.cgi?T=JS&NEWS=n&CSC=Y&PAGE=fulltext&D=ovft&AN=00003012-199902190-00008

Shin HY, Gerritsen ME, Bizios R (2002) Regulation of endothelial cell proliferation and apoptosis by cyclic pressure. Ann Biomed Eng 30(3):297–304. https://www.ncbi.nlm.nih.gov/pubmed/12051615. https://doi.org/10.1114/1.1458595

Shin HY, Bizios R, Gerritsen ME (2003) Cyclic pressure modulates endothelial barrier function. Endothelium (New York, N.Y.) 10(3):179–187. https://doi.org/10.1080/713715228

Shin HY, Underwood RM, Fannon MW (2012) Fluid pressure is a magnitude-dependent modulator of early endothelial tubulogenic activity: implications related to a potential tissue engineering control parameter. Tissue Eng Part A 18(ja):2590–2600. https://doi.org/10.1089/ten.TEA.2011.0588

Sikavitsas VI, Bancroft GN, Lemoine JJ, Liebschner MAK, Dauner M, Mikos AG (2005) Flow perfusion enhances the calcified matrix deposition of marrow stromal cells in biodegradable nonwoven fiber mesh scaffolds. Ann Biomed Eng 33(1):63–70. https://www.ncbi.nlm.nih.gov/pubmed/15709706. https://doi.org/10.1007/s10439-005-8963-x

Spatz JM, Wein MN, Gooi JH et al (2015) The wnt inhibitor sclerostin is up-regulated by mechanical unloading in osteocytes in vitro. J Biol Chem 290(27):16744–16758. https://doi.org/10.1074/jbc.M114.628313

Stanford CM, Morcuende JA, Brand RA (1995) Proliferative and phenotypic responses of bone-like cells to mechanical deformation. J Orthop Res 13(5):664–670. https://api.istex.fr/ark:/67375/WNG-0HN8FVJC-P/fulltext.pdf. https://doi.org/10.1002/jor.1100130505

Stavenschi E, Corrigan MA, Johnson GP, Riffault M, Hoey DA (2018) Physiological cyclic hydrostatic pressure induces osteogenic lineage commitment of human bone marrow stem cells: a systematic study. Stem Cell Res Ther 9(1):276. https://www.ncbi.nlm.nih.gov/pubmed/30359324. https://doi.org/10.1186/s13287-018-1025-8

Suda T, Takahashi N, Martin TJ (1992) Modulation of osteoclast differentiation. Endocr Rev 13(1):66–80. https://www.ncbi.nlm.nih.gov/pubmed/1555533. https://doi.org/10.1210/edrv-13-1-66

Takai E, Mauck RL, Hung CT, Guo XE (2004) Osteocyte viability and regulation of osteoblast function in a 3D trabecular bone explant under dynamic hydrostatic pressure. J Bone Miner Res 19(9):1403–1410. https://doi.org/10.1359/JBMR.040516

Tanaka SM, Sun HB, Roeder RK, Burr DB, Turner CH, Yokota H (2005) Osteoblast responses one hour after load-induced fluid flow in a three-dimensional porous matrix. Calcif Tissue Int 76(4):261–271. https://www.ncbi.nlm.nih.gov/pubmed/15812578. https://doi.org/10.1007/s00223-004-0238-2

Tani-Ishii N, Tsunoda A, Teranaka T, Umemoto T (1999) Autocrine regulation of osteoclast formation and bone resorption by IL-1α and TNFα. J Dent Res 78(10):1617–1623. https://doi.org/10.1177/00220345990780100601

Thompson GJ, Puleo DA (1996) Ti-6Al-4V ion solution inhibition of osteogenic cell phenotype as a function of differentiation timecourse in vitro. Biomaterials 17(20):1949–1954. https://doi.org/10.1016/0142-9612(96)00009-9

Torrance AG, Mosley JR, RFL S, Lanyon LE (1994) Noninvasive loading of the rat ulna in vivo induces a strain-related modeling response uncomplicated by trauma or periostal pressure. Calcif Tissue Int 54(3):241–247. https://www.ncbi.nlm.nih.gov/pubmed/8055374. https://doi.org/10.1007/BF00301686

Tsai M, Chen L, Zhou J et al (2009) Shear stress induces synthetic-to-contractile phenotypic modulation in smooth muscle cells via peroxisome proliferator-activated receptor α/δ activations by prostacyclin released by sheared endothelial cells. Circ Res 105(5):471–480. https://www.ncbi.nlm.nih.gov/pubmed/19628794. https://doi.org/10.1161/CIRCRESAHA.109.193656

Tsuzuki T, Okabe K, Kajiya H, Habu T (2000) Osmotic membrane stretch increases cytosolic ca(2+) and inhibits bone resorption activity in rat osteoclasts. Jpn J Physiol 50(1):67–76. https://www.ncbi.nlm.nih.gov/pubmed/10866699. https://doi.org/10.2170/jjphysiol.50.67

Turner CH (1992) Functional determinants of bone structure: beyond wolff's law of bone transformation. Bone (New York, N.Y.) 13(6):403–409. https://doi.org/10.1016/8756-3282(92)90082-8

Turner CH, Robling AG (2003) Designing exercise regimens to increase bone strength. Exerc Sport Sci Rev 31(1):45–50. https://www.ncbi.nlm.nih.gov/pubmed/12562170. https://doi.org/10.1097/00003677-200301000-00009

Turner CH, Forwood MR, Otter MW (1994) Mechanotransduction in bone: do bone cells act as sensors of fluid flow? FASEB J 8(11):875–878. http://www.fasebj.org/cgi/content/abstract/8/11/875. https://doi.org/10.1096/fasebj.8.11.8070637

Vadiakas GP, Banes AJ (1992) Verapamil decreases cyclic load-induced calcium incorporation in ROS 17/2.8 osteosarcoma cell cultures. Matrix (Stuttgart) 12(6):439–447. https://doi.org/10.1016/S0934-8832(11)80088-0

Wittkowske C, Reilly GC, Lacroix D, Perrault CM (2016) In vitro bone cell models: impact of fluid shear stress on bone formation. Front Bioeng Biotechnol 4:87. https://www.ncbi.nlm.nih.gov/pubmed/27896266. https://doi.org/10.3389/fbioe.2016.00087

Xing J, Li Y, Lin M et al (2014) Surface chemistry modulates osteoblasts sensitivity to low fluid shear stress. J Biomed Mater Res A 102(11):4151–4160. https://doi.org/10.1002/jbm.a.35087

Yoshino D, Funamoto K, Sato K, Kenry SM, Lim CT (2020) Hydrostatic pressure promotes endothelial tube formation through aquaporin 1 and ras-ERK signaling. Commun Biol 3(1):152. https://www.ncbi.nlm.nih.gov/pubmed/32242084. https://doi.org/10.1038/s42003-020-0881-9

You J, Reilly GC, Zhen X et al (2001) Osteopontin gene regulation by oscillatory fluid flow via intracellular calcium mobilization and activation of mitogen-activated protein kinase in MC3T3–E1 osteoblasts. J Biol Chem 276(16):13365–13371. https://www.ncbi.nlm.nih.gov/pubmed/11278573. https://doi.org/10.1074/jbc.M009846200

Yu W, Qu H, Hu G et al (2014) A microfluidic-based multi-shear device for investigating the effects of low fluid-induced stresses on osteoblasts. PLoS One 9(2):e89966. https://www.ncbi.nlm.nih.gov/pubmed/24587156. https://doi.org/10.1371/journal.pone.0089966

Zerwekh JE, Ruml LA, Gottschalk F, Pak CY, Blomqvist CG (1998) The effects of twelve weeks of bed rest on bone histology, biochemical markers of bone turnover, and calcium homeostasis in eleven normal subjects. J Bone Miner Res 13(10):1594–1601. https://ntrs.nasa.gov/citations/20040142203. https://doi.org/10.1359/jbmr.1998.13.10.1594

Zhao Y, Lv X, Liu Y et al (2015) Hydrostatic pressure promotes the proliferation and osteogenic/chondrogenic differentiation of mesenchymal stem cells: the roles of RhoA and Rac1. Stem Cell Res 14(3):283–296. https://doi.org/10.1016/j.scr.2015.02.006

Part III
In Vitro Studies of Stretch v. Hydrostatic Pressure

Chapter 3
Cellular Responses to Stretch Vs. Hydrostatic Pressure

Contents

J. Nagatomi, *Mechanobiology of Hydrostatic Pressure*, Frontiers of Biomechanics 5, https://doi.org/10.1007/978-981-95-4405-9_3

3.1 Cellular Responses to Stretch Vs Hydrostatic Pressure

As discussed earlier in Chap. 1, hollow organs like the heart, lungs, airway, blood vessels, and urinary bladder that contain fluids (air, blood, urine) are routinely subjected to changes in both HP and cyclic stretch simultaneously through daily physiological functions of those organs. In the past few decades, many researchers have examined the effects of cyclic stretching on cells from lungs and bronchi (epithelial cells, smooth muscle cells), and urinary bladder (urothelial cells, smooth muscle cells). To a lesser extent, investigators also examined the effects of physiological level HP on these cells. The differences in the amount of data available between cyclic stretch and HP studies are partly due to the availability of commercial devices that allow investigators to subject cultured to cells to controlled stretching. In this chapter, we will review the anatomy, physiology, and cellular composition to explore the biomechanical environment of the cells of these organs. We will then review the in vitro studies that aim to understand the role of cyclic stretch and HP in pathological conditions experienced by these organs.

3.1.1 Anatomy, Physiology, and Mechanical Loading Environments for Respiratory Cells In Vivo

Anatomically, the respiratory system consists of upper and lower respiratory tracts. Functionally, it can be divided into a portion that transports air (e.g. nose, nasal cavity, pharynx in the upper tract, and larynx, trachea, bronchi in the lower tract) and a portion where gas exchange takes place (respiratory bronchioles, alveolar ducts, and alveoli of the lungs) (McKinley and O'Loughlin 2005). The types of cells that are mainly of interest to mechanobiology research are the epithelial cells that line parts of airways and alveoli, and the smooth muscle cells that form the walls of bronchioles because they are subjected to various mechanical forces physiologically and pathologically.

Pulmonary ventilation, which means flow of air in and out of the lungs, can take place by motion of the diaphragm muscles to expand and contract the lungs, or by using a mechanical ventilator when a person is not able to breath on their own. Thus, the cells of the respiratory system are constantly subjected to dynamic changes in pressure and stretch. However, the stretching of the cells can become excessive under forced mechanical ventilation, which leading to various injuries of the airways. For these reasons, numerous in vitro studies have been conducted in the past few decades to elucidate the link between mechanical loading such as cyclic stretching and the function of epithelial and smooth muscle cells of the respiratory system.

3.1.2 *Effects of Cyclic Stretching on the Cells of the Respiratory System*

Breathing is essential for living and the cyclic motion of the lungs continues throughout our lives. Thus, the cells of the respiratory system are designed to withstand repeated stretching and pressure changes without breaking. However, mechanical ventilation, a form of life support that is need when a person is having a surgery or cannot breathe on their own due to an illness, can cause or aggravate lung and airway injuries. For this reason, numerous in vitro studies using alveolar epithelial cell models have been conducted to elucidate the cellular and molecular mechanisms underlying these so-called Ventilator-Induced Lung Injuries (VILI). Moreover, independent from VILI, some of the investigators that are studying the disease mechanisms of asthma have recognized various effects of cyclic stretching on the function of airway smooth muscle cells. Below are short reviews of studies examining the effects of cyclic stretching on lung (alveolar) epithelial cells, and on airway (bronchial) smooth muscle cells.

The main goal of studying the effect of cyclic stretching on alveolar epithelial cells is to understand the direct role of cell deformation in development of VLLI and significant advances have been made thanks to biomechanics and mechanobiology researchers in the past few decades. Early studies demonstrated that cell deformation caused increases in surfactant releases (Wirtz and Dobbs 1990), cell proliferation (Liu et al. 1992) and prostacyclin production (Skinner et al. 1992) in alveolar epithelial cells. While these studies provided valuable data, it was pointed out that the experimental setups used in these studies had some limitations in simulating the deformation experienced by alveolar epithelial cells in vivo (Tschumperlin and Margulies 1998). With more sophisticated analysis of the mechanical environment of the lung and development of experimental devices (Tschumperlin and Margulies 1998; Davidovich et al. 2013), investigators have made many critical discoveries in the relationship between cell deformation and alveolar epithelial cell injury (Tschumperlin et al. 2000; Fisher and Margulies 2002; Cohen et al. 2010; Cavanaugh et al. 2006). More specifically, it was shown that deformation-induced alveolar epithelial cell injury was dependent on stretch magnitudes (Tschumperlin and

Margulies 1998), duration, and frequency of the repetitive stretch (Tschumperlin et al. 2000). Moreover, exposure of alveolar epithelial cells to well-controlled cyclic (0.25 Hz) stretch (up to 37%) resulted in increased activities of the membrane-bound sodium ion pump (Fisher and Margulies 2002) and increased transepithelial permeability (Cavanaugh et al. 2006), which was mediated by phosphorylation of MAPK such as JNK and ERK (Cohen et al. 2010). Together, these and a number of other findings made significant contributions toward improved understanding of the role of cyclic stretching in VLLI.

The rationale behind studies looking at the effects of stretching on airway smooth muscle (ASM) cells in relation to asthma pathology is that these cells are constantly subjected to approximately 3% cyclic stretching in vivo from normal breathing (Maksym et al. 2005). Investigators consider that it is necessary for in vitro cell cultures to be subjected to cyclic stretching to better understand the pathophysiology of ASM. For example, asthma is characterized by inflammatory reactions and airway remodeling caused by smooth muscle hyperplasia, increased collagen deposition (Bonacci et al. 2003), elevated contractile tone and cytoskeletal changes associated with cell stiffening (Maksym et al. 2005). Thus, several in vitro studies examined the effects of stretching on ASM cell proliferation (Bonacci et al. 2003; Smith et al. 1994; Hasaneen et al. 2005) and remodeling of cytoskeletal filaments (Smith et al. 1997; Smith et al. 2003; Deng et al. 2004; Rosenblatt et al. 2007) with some mixed results. More specifically, when ASM cells were exposed to relatively high level of stretch (15 ~ 30%) cell proliferation increased (Smith et al. 1994; Hasaneen et al. 2005), which was shown in a recent study that mediated by overexpression of microRNA miR-10b-5p (Quan et al. 2022). In contrast, when ASM cells were subjected to a more physiological level of stretch (4%), cell proliferation was lower when compared to cells that were cultured under static conditions (Bonacci et al. 2003). These results demonstrate the bimodal effects of cyclic stretch on ASM cell proliferation: while the extreme level of stretching to model bronchospasm/hyperinflation (Hasaneen et al. 2005) and removal of stretch to model asthmatic airway (Bonacci and Stewart 2006) can both create a similar increase in cell proliferation suggesting that high levels of stretching can be harmful while physiological level stretching is beneficial. The harmful effect of large strain has also been observed in terms of cell stiffening. Specifically, AMS cells exhibited increases in stiffening when exposed to substrate stretching up to 20% (Rosenblatt et al. 2007) and when cytoskeletal filaments were directly stimulated via magnetic twisting (Deng et al. 2004) indicating that the ASM cytoskeleton may be the stretch sensing element that contributes to pathological responses of these cells such as hyperplasia and stiffening.

3.1.3 Effects of HP on Respiratory Cells

In addition to excessive stretching, under asthmatic bronchoconstriction, the pressure environments of the cells of the respiratory system can alter significantly. Transmural pressure, which is the hydrostatic pressure (HP) differences between the lumen of the pulmonary organs (airway and lungs) and the pleural cavity ($\Delta P = P_{lumen} - P_{pleural}$) is considered a critical mechanical parameter for both development of the organs (Nelson et al. 2017; Jaslove et al. 2022) and in pathological conditions such as fibrosis due to chronic asthma (Tschumperlin et al. 2003).

To elucidate the mechanism for airway remodeling due to asthmatic bronchoconstriction, a series of studies were conducted where human bronchial epithelial cells were exposed to simulated pathological level transmural pressures (~30 cmH_2O) in vitro (Tschumperlin et al. 2003; Tschumperlin et al. 2004; Chu et al. 2005; Tschumperlin et al. 2002). It was found that application compressive stress due to transcellular pressure gradient across a monolayer of bronchial epithelial cells led to rapid activation of extracellular signal-regulated kinase (ERK) and heparin-binding epidermal growth factor-like growth factor (HB-EGF) expression (Tschumperlin et al. 2002). The proposed mechanism for this mechanically triggered, ERK-dependent increases in HB-EGF and other EGF receptor family (EGFR) ligands (Chu et al. 2005) involves collapse of the intercellular space between adjacent epithelial cells and increases in EGFR ligands (Tschumperlin et al. 2004). These ligands stimulate both ERK activation and EGFR, as well as can feedback their own receptors in an autocrine or paracrine manner. These bronchial epithelial cell responses to high transmural pressures (~30 cmH_2O) have been linked to increased release of fibrotic mediators (Tschumperlin et al. 2003), which can affect subepithelial cells to elicit matrix remodeling (Swartz et al. 2001) characteristic of chronic asthma.

In contrast to the pathological level transmural pressure that is considered harmful to airway cells, the physiological level pressure (300 Pa or ~ 3 cmH_2O) is considered necessary for proper development and morphogenesis of the respiratory system (Nelson et al. 2017). Although the link between reduced transmural pressure and maldevelopment of the lungs was observed several decades ago (Fewell et al. 1983; Harding and Hooper 1996), it was only recent, with more sophisticated experimental approaches, that the quantitative relationship between transmural pressure and fetal lung development was experimentally observed (Nelson et al. 2017). Early studies demonstrated that the fluid secreted by epithelial cells and contraction of smooth muscle generate the positive transmural pressure in the order of 2 ~ 4 cmH_2O (Schittny et al. 2000). When mouse embryonic (E12) lung explants were subjected to transmural pressure (ΔP = 20 ~ 300 Pa, 0.2 ~ 3.0 cmH_2O for 48 h) using a custom microfluidic chamber, lung development occurred in pressure-dependent manner (Nelson et al. 2017). More specifically, with higher ΔP epithelial branching increased and airway smooth muscle contraction decreased, indicating that transmural pressure regulates the epithelial function and cross-talk with subepithelial tissue to control the development of the lungs (Nelson et al. 2017).

3.1.4 Anatomy, Physiology, and Mechanical Loading Environments for Urinary Tract Cells In Vivo

Like the respiratory system, the urinary tract is also divided into two major sections, the upper urinary tract that includes the kidneys and ureters, and the lower urinary tract that includes the bladder and urethra. The kidney's function is to filter the circulating blood to remove waste and excess fluid, which become urine to be transported through the ureters. The function of the bladder, on the other hand, is to store the urine and void it periodically by coordinated actions of contracting bladder smooth muscle and relaxing urethral sphincter muscle. Because the tissues of the bladder are constantly subjected to the storage-voiding cycles and the associated mechanical loading stimuli, mechanobiology research involving cyclic stretch and HP has mainly focused the bladder cells.

The bladder wall tissue consists of four major layers, the inner most urothelium that is separated by the lamina propria from the detrusor muscle, and the surrounding connective tissue-rich serosa/adventitia (McKinley and O'Loughlin 2005). The urothelium is composed of three layers of urothelial cells, apical, intermediate and basal layers that are also known as transitional epithelium because of its ability to transition from 5–7 layer thick in an empty bladder to 2–3 layers in a full bladder. The detrusor muscle, especially near the bladder neck, is also composed of three layers of highly oriented smooth muscle cells, inner longitudinal, middle circular, and outer longitudinal layers although the rest of the bladder contains randomly oriented smooth muscle cells (McKinley and O'Loughlin 2005). During the urine storage phase the bladder wall tissue experiences gradual increases in distension, but low intravesical pressure (~5 cmH_2O) is maintained by smooth muscle relaxation. Sensing that the bladder is full and it needs to be emptied is an important physiological function. Conventionally, it was understood that the mechanosensitive afferent that innervates the bladder muscle sense the tension in the bladder wall when it is full (de Groat and Yoshimura 2009). However, more recent studies suggest that the urothelium that is being stretch during filling is also involved in the sensory function via activation of membrane-bound ion channels (Janssen et al. 2017). Under clinical conditions such as acute urinary retention (AUR) or bladder outlet obstruction (BOO) due to enlarged prostate in men, inability to void puts the bladder wall tissue for a long period of stretch. These changes in the loading environment such as excessive distention among many voiding dysfunctions may be triggering pathological tissue remodeling such as smooth muscle hypertrophy and fibrosis (Fusco et al. 2018). To improve understanding of bladder physiology and pathology, numerous in vitro studies have examined the effects of stretch on both bladder urothelial and smooth muscle cells.

3.1.5 *Effects of Stretch on Urinary Tract Cells*

To study the effect of stretch on bladder urothelial or smooth muscle cells in vitro, the majority of studies rely on commercial devices such as Flexcell system (Flexcell International, Burlington, NC, USA) (Sun et al. 2001; Sun et al. 2002; Adam et al. 2002) to subject cells to cyclic stretch (e.g. 20% for 1 s followed by 2 s rest) for a long period of time (24–96 h), or Strex system (STREX Inc., Osaka, Japan) (Ihara et al. 2018; Mochizuki et al. 2009) to subject cells to a single episode of quick stretch (10–30% in 1 s). Both of these and other similar systems use flexible substrates on which cells are seeded as a monolayer and stretching is achieved either by applying negative pressure on the bottom side of the substrate for biaxial stretch (Flexcell) or by using a linear actuator to apply uniaxial stretch (Strex), respectively. Although it is debatable whether these cell stretching regimens are accurate representation of the mechanical loading conditions of the bladder in vivo, following exposure of urothelial or smooth muscle cells to stretch, the investigators examine the cellular and molecular responses that are pertinent to the bladder pathophysiology.

As mentioned previously, it has been agreed by many investigators that one of the roles of the bladder urothelial cells is to sense the bladder fullness (Janssen et al. 2017; Birder 2005). Numerous studies reported that stretching of urothelial cells using both isolated bladders and cultured primary urothelial cells lead to releasing adenosine triphosphate (ATP), which acts as a signaling molecule for afferent nerves and also for urothelial cells themselves (Sun et al. 2001; Birder 2005; Negoro et al. 2014; Sellers et al. 2018). For example, Sun et al. conducted a series of studies to test a hypothesis that various forms of lower urinary tract conditions such as urinary frequency and urgency result from increased sensitivity of the urothelial cells to stretch (Sun et al. 2001; Sun et al. 2002). Specifically, when urothelial cells harvested from patients of interstitial cystitis (Sun et al. 2001) or benign prostate hyperplasia (BPH) (Sun et al. 2002) were subjected to cyclic biaxial stretch (20%, 1 s on 2 s off) for up to 96 h, their ATP release was significantly greater compared to that of normal urothelial cells, indicating that the pathological group cells were more sensitive to the stretch stimuli. To determine how urothelial cells sense stretch stimuli, another group conducted studies focusing on membrane bound mechanosensitive ion channels, TRPV4 (Mochizuki et al. 2009) and Piezo1 (Miyamoto et al. 2014). These studies using urothelial cells harvested from knock out mice for TRPV4 and siRNA knock down of Piezo1 provided evidence that, in response to a single episode of uniaxial stretch (~17.5% in 1 s), both ATP release and intracellular Ca^{2+} increase were attenuated when compared to those of wild type cells (Miyamoto et al. 2014). The authors concluded that stretching of urothelial cells lead to activation of both TRPV4 and Piezo1 channels independently, which allow entry of Ca^{2+} ions into the cells needed for subsequent ATP release (Miyamoto et al. 2014).

As mentioned above, one of the clinical motivations to study the effects of stretch on bladder smooth muscle cells is to understand the cellular and molecular mechanism for the pathological tissue remodeling due to bladder outlet obstruction. To model the mechanical changes due to bladder outlet obstruction, these studies subjected smooth muscle cells cultured on deformable membrane to cyclic stretching at frequencies of 5–20 cycles per min (Park et al. 1999a; Park et al. 1998; Steers et al. 1998; Chaqour et al. 2002; Yu et al. 2003; Adam et al. 2004; Dai et al. 2015). It should be noted, however, that the most notable change in the mechanical environment for the bladder with obstructed outlet is the abnormally high intravesical pressure during voiding (Yamaguchi 2004). This high pressure is generated by contraction of the detrusor smooth muscle cells and not by an external pressure source that subjects smooth muscles to excessive stretch. In addition, voluntary voiding of the bladder occurs every so many hours, and not every so many seconds. However, in the past two decades the high-frequency cyclic stretching of bladder smooth muscle cells using commercial setups such as Flexcell devices has been widely accepted as the model in numerous studies to represent the altered mechanical environment for the smooth muscle cells due to bladder outlet obstruction. The results of these studies provided evidence that exposure of bladder smooth muscle cells to cyclic stretching led to increased cell proliferation (Park et al. 1999a; Dai et al. 2015) and increased release of various signaling molecules such as HB-EGF (Park et al. 1998) and COX-2 (Park et al. 1999a) as well as changes in gene expression (Chaqour et al. 2002; Adam et al. 2004) that associated with cell proliferation. Together, the authors of these and other in vitro studies conclude that tissue remodeling such as smooth muscle thickening associated with bladder outlet obstruction results from stretch-activated cell signaling pathways that lead to increases in smooth muscle cell proliferation (Yamaguchi 2004).

3.1.6 Effects of HP on Urinary Tract Cells

Under the pathological conditions known as bladder outlet obstruction (BOO), the bladder cells are subjected to not only prolonged stretch due to incomplete emptying, but also high intravesical pressure due to voiding against high resistance outlet. Thus, some investigators have focused on the effect of elevated HP, instead of increased stretching, on bladder tissue remodeling and investigated using various in vitro models (Fusco et al. 2018; Haberstroh et al. 2000; Haberstroh et al. 1999; Backhaus et al. 2002) Specifically, earlier work by Haberstroh et al. used hydrostatic columns of culture media to expose monolayer cultures of bladder smooth muscle cells to elevated HP (4–8.5 cmH_2O above atmospheric pressure) (Haberstroh et al. 2000; Haberstroh et al. 1999). The results of these studies demonstrated that bladder smooth muscle cell proliferation increased under elevated pressure and that the pressure-induced cell proliferation was mediated by soluble mitogenic factors released by these cells in response to the elevated pressure (Haberstroh et al. 1999). A subsequent study by these authors used a more sophisticated, computer-controlled experimental setup to expose bladder smooth muscle cell cultures to elevated HP (up to 40 cmH_2O) and demonstrated that HP treatment led to decreases in MMP-1, 2, and 9, and increases in TIMP-1 levels (Backhaus et al. 2002). Together, these studies provided evidence that bladder smooth muscle cells are sensitive to elevated physiological level HP, which can trigger cellular responses related to bladder tissue remodeling.

3.2 Cyclic Pressure Stimulates DNA Synthesis through the PI3K/Akt Signaling Pathway in Rat Bladder Smooth Muscle Cells

3.2.1 Introduction

Urinary tract dysfunctions, such as those associated with spinal cord injury (SCI), can create abnormal mechanical stresses on bladder cells. In SCI patients, impaired coordination between the bladder and external sphincter often results in incomplete emptying of the bladder (De Groat 1995). This dysfunction produces chronically elevated intravesical pressure, which, over time, is accompanied by tissue-level alterations including thickening of the bladder wall and fibrosis (Ewalt et al. 1992). The detrusor muscle typically exhibits smooth muscle hypertrophy and/or hyperplasia, along with increased extracellular matrix (ECM) protein deposition (Karim et al. 1993; Peters et al. 1992). These changes contribute to reduced compliance, making the bladder stiffer and functionally a high-pressure reservoir, thereby increasing the risk of upper urinary tract and renal damage (De Groat 1995).

Previous in vitro studies demonstrated that bladder smooth muscle cells (SMCs) respond to cyclic stretch by undergoing hyperplasia (Galvin et al. 2002; Orsola et al. 2002) and upregulating growth factors such as heparin-binding EGF-like growth factor (HB-EGF) and enzymes like cyclooxygenase-2 (COX-2) (Park et al. 1999a; Nguyen et al. 2000; Park et al. 1999b). These mechanically induced responses were associated with activation of several intracellular signaling cascades, including MAPK, ERK, and PI3K/Akt pathways (Adam et al. 2003; Nguyen et al. 2000). While cyclic stretch has been used extensively to model mechanical stimulation of bladder cells, it only partially represents physiological and pathological loading conditions. Under normal filling, bladder pressure remains relatively low. However, when the bladder is near capacity—particularly in obstructed or SCI-affected bladders—contraction of the smooth muscle can markedly elevate intravesical pressure without further stretching the wall. In such cases, bladder cells are subjected to oscillating hydrostatic pressure rather than distension.

On this basis, it was hypothesized that cyclic hydrostatic pressure may act as a key signal contributing to bladder remodeling following SCI. To address this, the present study examined how cyclic pressure influences DNA synthesis and signaling activity in rat bladder smooth muscle cells, using an in vitro experimental model.

3.2.2 *Summary of Experimental Results*

Primary bladder smooth muscle cells (SMCs) were isolated from female Sprague–Dawley rats following published procedures (Kropp et al. 1999) and expanded in RPMI medium supplemented with 10% FBS. For pressure experiments, cultures were maintained in either 1% or 10% FBS conditions. Cells were then placed in a custom-built cyclic pressure system (Nagatomi et al. 2001) and subjected to sinusoidal oscillations ranging from 0 to 40 cmH_2O at frequencies of 0.1 Hz or 0.02 Hz, for durations up to 24 h. Static cultures maintained under identical conditions served as controls. Medium volume remained stable throughout the experiments, and measurements of pH, pCO_2, and pO_2 in pressurized cultures were indistinguishable from static controls at all time points (Fig. 3.1), confirming that the observed cellular responses were attributable to pressure rather than altered culture conditions.

After 24 h, total DNA content of SMCs exposed to 0.02 Hz cyclic pressure was not significantly different from that of static controls (Fig. 3.2). By contrast, cells exposed to 0.1 Hz cyclic pressure showed a significant increase in DNA content compared with static controls under both 1% FBS ($p = 0.004$) and 10% FBS ($p = 0.008$). Serum concentration alone (1% vs. 10%) did not alter DNA levels ($p = 0.526$), nor was there an interaction between serum concentration and pressure conditions ($p = 0.749$). TUNEL staining revealed very few apoptotic nuclei, with no detectable differences between pressurized and static groups (results not shown). BrdU incorporation assays demonstrated significantly greater DNA synthesis in SMCs subjected to cyclic pressure compared with static controls, both in 1% FBS ($p = 0.007$) and 10% FBS ($p = 0.0009$) conditions (Fig. 3.3). Together, these data support that cyclic pressure stimulates proliferation rather than cell death.

Levels of phosphorylated Akt (Ser473) and total Akt were quantified after 10, 30, 60 min, and 12 h of cyclic pressure exposure. Akt phosphorylation did not differ from static controls at 10 min ($p = 0.641$) or 30 min ($p = 0.646$), but was significantly elevated after 1 h ($p < 0.001$) and remained elevated at 12 h ($p < 0.05$; Fig. 3.4). To determine whether the PI3K/Akt pathway mediated the proliferative effects of pressure, cells were pretreated with either wortmannin (50 nM) or LY294002 (20 μM). Under static conditions, BrdU incorporation remained unchanged in the presence or absence of inhibitors ($p = 0.129$). However, in cultures exposed to cyclic pressure, BrdU incorporation was significantly suppressed by both wortmannin ($p < 0.05$) and LY294002 ($p = 0.0005$; Fig. 3.4). Taken together, these experiments show that cyclic hydrostatic pressure promotes proliferation of bladder SMCs in a frequency-dependent manner, and that this effect is mediated through activation of the PI3K/Akt signaling pathway.

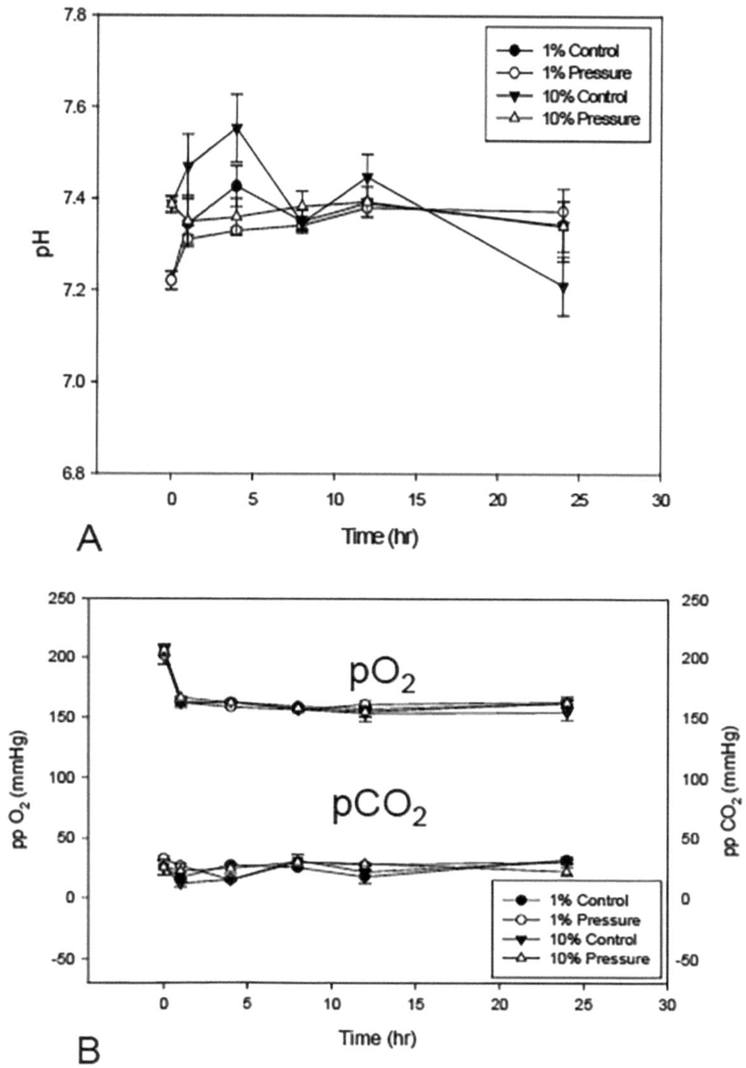

Fig. 3.1 Stability of culture media under cyclic pressure. Measurements of pH (**a**), pCO_2, and pO_2 (**b**) showed no differences between media exposed to cyclic pressure and static controls, under either 1% or 10% serum conditions, across all tested time points. (Adapted with permission from *Ann Biomed Eng*, Springer (2007))

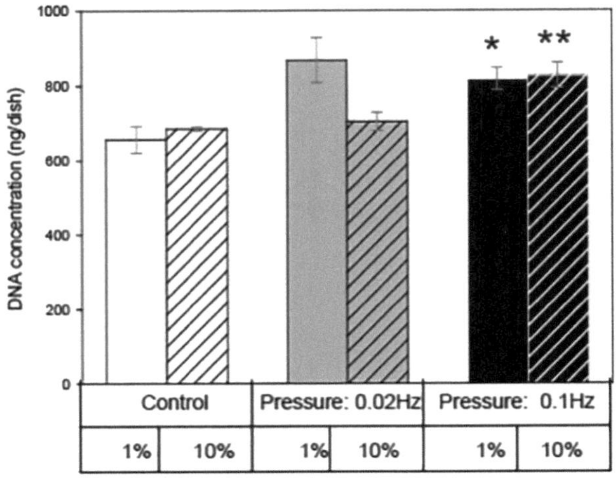

Fig. 3.2 DNA content in rat bladder smooth muscle cells after cyclic pressure exposure. DNA concentration was unaffected by serum concentration (1% vs. 10%) in both static ($p = 0.502$, $n = 3$) and pressure ($p = 0.820$, $n = 3$) groups. At 0.02 Hz, DNA levels were similar to static controls ($p = 0.352$, $n = 3$). However, at 0.1 Hz, SMCs showed significantly greater DNA concentration than static controls in both serum conditions. Data are mean ± SEM; *p = 0.04, **p = 0.008, n = 3. (Adapted with permission from *Ann Biomed Eng*, Springer (2007))

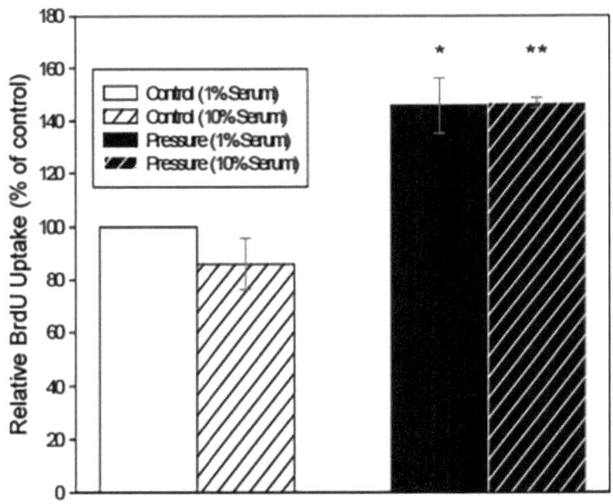

Fig. 3.3 DNA synthesis in rat bladder SMCs under cyclic pressure. BrdU incorporation was similar across 1% and 10% serum conditions in static cultures ($p = 0.234$, $n = 3$) and in pressurized cells ($p = 0.822$, $n = 3$). However, 0.1 Hz cyclic pressure significantly increased BrdU uptake compared with static controls under both serum conditions. Data are mean ± SEM; *p = 0.007, **p = 0.005, n = 3, relative to 1% serum static control. (Adapted with permission from *Ann Biomed Eng*, Springer (2007))

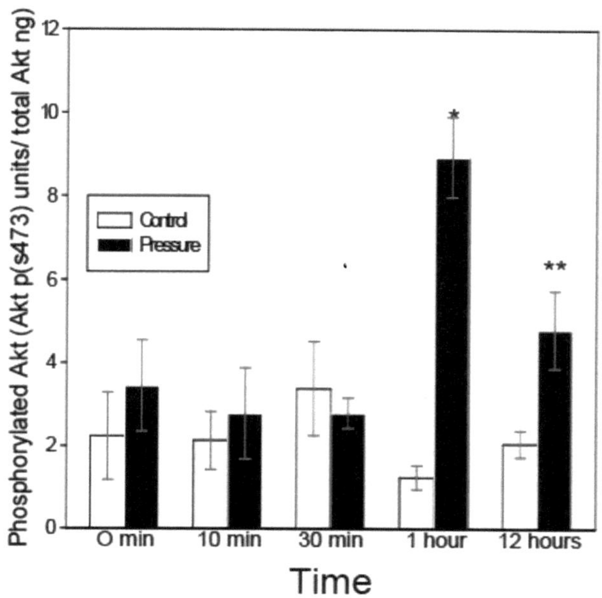

Fig. 3.4 Akt phosphorylation in SMCs subjected to cyclic pressure. Akt phosphorylation was unchanged in static cultures ($p = 0.474$, $n = 3$) and in pressurized cells at 10 min ($p = 0.641$, $n = 3$) or 30 min ($p = 0.646$, $n = 3$). By contrast, phosphorylation was significantly increased after 1 h ($p = 0.002$, $n = 3$) and remained elevated at 12 h ($p < 0.05$, $n = 3$). Data are mean ± SEM. (Adapted with permission from *Ann Biomed Eng*, Springer (2007))

3.2.3 Discussion

This study is the first to examine cyclic hydrostatic pressure as a mechanostimulatory signal capable of altering smooth muscle cell (SMC) behavior relevant to bladder remodeling. In these experiments, cultured bladder SMCs were exposed to oscillatory pressure waves between 0 and 40 cmH_2O at either 0.1 Hz or 0.02 Hz. These conditions were chosen to approximate the non-voiding contractions observed after spinal cord injury, where rodent models have shown mean contraction pressures of 28–45 cmH_2O with a frequency of about one contraction every 10 seconds (0.1 Hz) (McGuire et al. 2002; Abdel-Karim et al. 2002; Seki et al. 2002). Thus, the protocol simulated isovolumetric contractions, which elevate bladder wall stress in the absence of wall stretch, as occurs during obstruction.

When SMCs were exposed to 0.1 Hz cyclic pressure for 24 h, DNA content was significantly elevated relative to static controls under both 1% and 10% serum conditions (Fig. 3.2). Because serum factors are known to promote growth, experiments were performed under both low and normal serum conditions to determine whether serum masked the pressure response. Statistical analysis demonstrated that the proliferative effect was attributable solely to pressure, independent of serum level. Additionally, pH, pCO_2, and pO_2 of the media remained unchanged compared to static controls (Fig. 3.1), ensuring that the observed proliferative effect was directly due to cyclic pressure rather than altered gas exchange.

To distinguish whether the increase in DNA content was the result of proliferation or reduced apoptosis, both TUNEL assays and BrdU incorporation were performed. TUNEL staining showed minimal apoptosis across all conditions, with no difference between cyclic pressure and static controls. In contrast, BrdU labeling revealed significantly increased DNA synthesis in SMCs subjected to cyclic pressure (Fig. 3.3). Together, these results confirm that cyclic pressure enhances proliferation rather than reducing cell death

To identify signaling pathways mediating this effect, phosphorylation of Akt, a key regulator of survival, hypertrophy, and proliferation (Adam et al. 2003; Brazil and Hemmings 2001), was examined. Akt phosphorylation was significantly upregulated after 1 h ($p < 0.001$) and remained elevated at 12 h ($p < 0.05$) under cyclic pressure compared to static controls (Fig. 3.4). Importantly, the pressure-induced increase in DNA synthesis was abolished by pharmacological blockade of PI3K/Akt using either wortmannin (50 nM) or LY294002 (20 μM) (Fig. 3.5). Since wortmannin can influence off-target pathways (Cross et al. 1995), inclusion of LY294002 further validated the central role of PI3K/Akt in this proliferative response

Akt is known to be activated by a variety of upstream inputs, including growth factors (Adam et al. 2003), GPCR signaling (Stoyanov et al. 1995), and elevations in intracellular Ca^{2+} (Danciu et al. 2003). Conditioned medium collected from pressurized SMC cultures failed to induce proliferation when applied to untreated cells

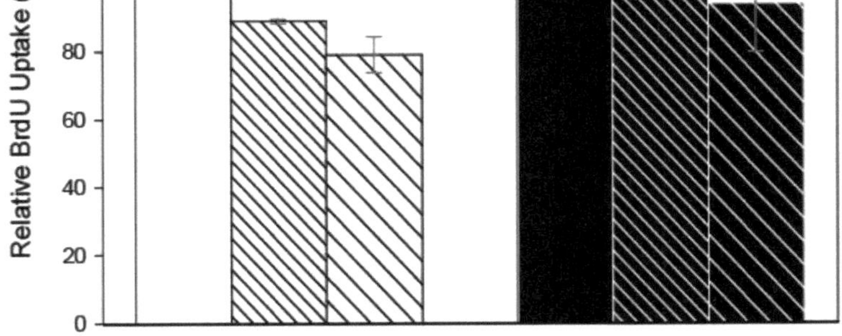

Fig. 3.5 PI3K/Akt dependence of cyclic pressure–induced DNA synthesis. DNA synthesis was similar across all static groups regardless of inhibitor treatment ($p = 0.129$, $n = 3$). Under cyclic pressure, however, BrdU incorporation was significantly reduced in the presence of wortmannin or LY294002, compared to pressurized cells without inhibitors. Data are mean ± SEM; *$p < 0.05$, **$p = 0.0005$, $n = 3$. (Adapted with permission from *Ann Biomed Eng*, Springer (2007))

(data not shown), indicating that soluble paracrine factors were not responsible for Akt activation in this model. One plausible explanation is direct activation of G protein–coupled receptors, as seen in other systems. For instance, in a mouse model of cardiac pressure overload induced by transverse aortic constriction, Akt phosphorylation and cardiac hypertrophy were absent in Gq-null transgenic animals, implicating G-protein–mediated signaling as a key mechanism (Bommakanti et al. 2000; Kim et al. 2004; Naga Prasad et al. 2000). Another possible pathway involves Ca^{2+}-dependent PI3K activation, which has been documented in cells exposed to mechanical strain (Danciu et al. 2003; Farrugia et al. 1999; Vadiakas and Banes 1992). Although the current model applied hydrostatic pressure without stretch, it is conceivable that pressure-sensitive ion channels facilitated intracellular Ca^{2+} influx, leading to Akt activation. Additional studies will be required to test this hypothesis and clarify the precise mechanotransduction pathway.

3.2.4 Conclusions

Chronic exposure of rat bladder SMCs to cyclic hydrostatic pressure in vitro resulted in enhanced proliferation mediated by the PI3K/Akt pathway. These findings suggest that non-voiding contractions, which subject the bladder to oscillatory pressure, may act as a mechanobiological trigger for bladder wall remodeling and fibrosis in pathological conditions such as spinal cord injury and outlet obstruction.

3.3 The Effect of Elevated Hydrostatic Pressure on Rat Bladder Smooth Muscle Cell Morphology and Proliferation

3.3.1 Introduction

Cells within the lower urinary tract are constantly exposed to mechanical forces, particularly hydrostatic pressure and stretch. Under normal conditions these forces are critical for bladder function; however, abnormal mechanical environments, such as those associated with spinal cord injury (SCI) (Korossis et al. 2006; Watanabe et al. 1996) or bladder outlet obstruction (BOO) (Chancellor et al. 1994), can disrupt cellular behavior and compromise bladder integrity. Such disruptions contribute to smooth muscle hypertrophy, hyperplasia, and fibrosis, ultimately leading to altered tissue architecture and impaired compliance of the bladder wall (Korossis et al. 2006).

In vitro culture systems have been widely used to investigate how bladder smooth muscle cells (BSMCs) respond to mechanical stimulation. For instance, cyclic stretch (20% strain, 0.1 Hz) applied for up to 48 h has been shown to increase

BSMC size (Karim et al. 1993; Peters et al. 1992), cell number (Adam et al. 2002; Adam et al. 2003), and to regulate stretch-sensitive genes, including heparin-binding EGF-like growth factor (HB-EGF) (Nguyen et al. 2000) and the cysteine-rich protein Cyr61 (Tamura et al. 2001). In addition to stretch, bladder cells are also subjected to transmural hydrostatic pressure. A seminal study by Haberstroh et al. demonstrated that ovine BSMCs exposed to elevated static pressures (4–8.5 cmH_2O) for up to 7 days exhibited significantly increased proliferation after 5 days, which could be suppressed with an HB-EGF inhibitor (CRM197) (Haberstroh et al. 2000; Haberstroh et al. 1999).

It is well established that smooth muscle cells (SMCs) can undergo phenotypic modulation between contractile and synthetic states. Compared with contractile SMCs, which are elongated and express smooth muscle-specific proteins, synthetic SMCs are characterized by reduced expression of these proteins (Roby et al. 2008; Worth et al. 2001), increased proliferation (Roby et al. 2008; Worth et al. 2001; Sobue et al. 1998; Stegemann and Nerem 2003), remodeling of the extracellular matrix (ECM) (Backhaus et al. 2002; Worth et al. 2001; Adelöw and Frey 2007), and a more spread morphology (Roby et al. 2008; Rensen et al. 2007). These phenotypic changes are believed to play an important role in long-term bladder remodeling under pathological conditions.

Based on this evidence, we hypothesized that BSMCs exposed to elevated hydrostatic pressure may exhibit both morphological and functional changes consistent with a phenotypic shift. To test this, the present study employed an in vitro model of sustained hydrostatic pressure and evaluated its effects on cell morphology, ERK½ signaling, proliferation, and expression of contractile proteins. Since signaling pathways such as PI3K/Akt (Adam et al. 2003), p38 (Nguyen et al. 2000), and ERK½ (Halachmi et al. 2006) have previously been implicated in stretch-induced cellular responses, we also sought to determine whether ERK½ activation mediates BSMC responses to hydrostatic pressure.

3.3.2 *Summary of Experimental Results*

Primary bladder smooth muscle cells (BSMCs) were isolated from female Sprague–Dawley rats following established protocols (Kropp et al. 1999) and maintained in RPMI medium supplemented with 10% FBS under standard conditions (37 °C, humidified 95% air/5% CO_2). To test the effects of mechanical loading, cells were subjected to sustained hydrostatic pressure of 7.5 cmH_2O above atmospheric pressure using a custom system described previously (Haberstroh et al. 2000; Haberstroh et al. 1999). Exposure times ranged up to 48 h, with cells cultured under atmospheric pressure serving as controls. In selected experiments, cells were pretreated with 10 μM U0126 (a MEK½ inhibitor) for 2 h prior to pressure application.

When exposed to pressure for 4 h, BSMCs underwent a noticeable morphological transition from elongated, spindle-like shapes toward a more rounded appearance (Fig. 3.6a, b). Morphology was quantified by calculating aspect ratios, which

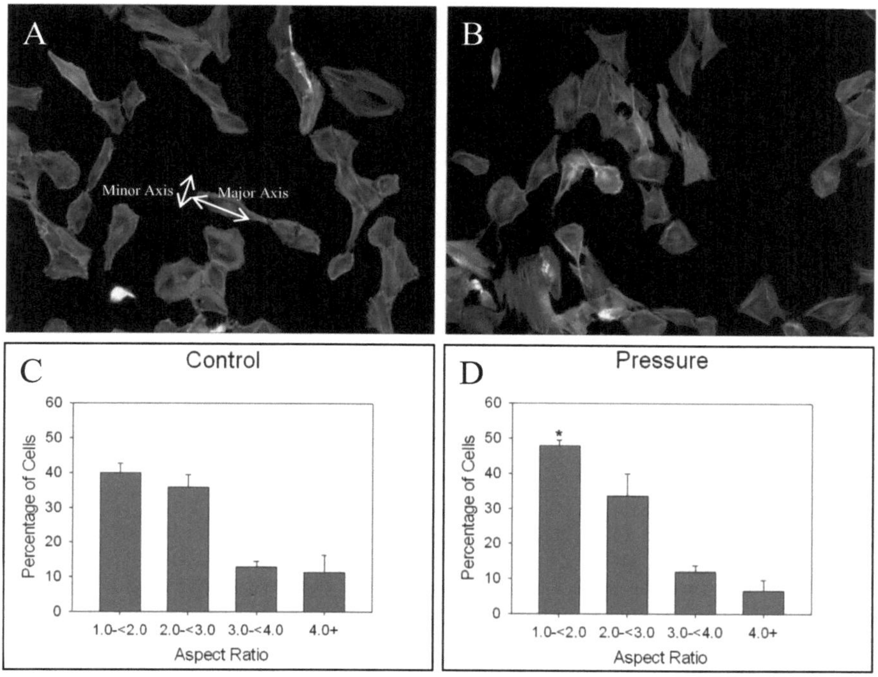

Fig. 3.6 Morphological changes of rat BSMCs under sustained hydrostatic pressure. Primary BSMCs were subjected to 7.5 cm H_2O of sustained pressure for 4 h, fixed, and stained with rhodamine-phalloidin (actin filaments) and DAPI (nuclei). [Top] Representative images of control cells (**a**) and cells after pressure exposure (**b**). Arrows indicate the major and minor axes used for aspect ratio calculations (magnification 100×). [Bottom] Histograms of aspect ratios from control cells (**c**) and pressure-exposed cells (**d**). Exposure to pressure increased the proportion of rounded cells (aspect ratio 1–2) and reduced the fraction of elongated cells (aspect ratio $\geq$ 4). Data are mean ± SD; analyzed with Student's t-test; $^*p < 0.05$; n = 3 with 384 cells/group per experiment. (Adapted with permission from *Urology*, Elsevier (2010))

were displayed as histograms (Fig. 3.6c, d). Compared to controls, pressure-treated cells showed a marked increase in the fraction of cells with aspect ratios between 1.0–2.0 (rounded morphology) and a corresponding reduction in those with aspect ratios ≥4.0 (elongated morphology).

To assess signaling responses, cells were harvested at 30, 90, and 180 min for evaluation of ERK½ phosphorylation, and at 24 and 48 h for phenotypic protein expression. Western blotting was performed using antibodies against α-SMA, SM22, GAPDH, total ERK½, and phosphorylated ERK½. Densitometric analysis of chemiluminescence images revealed that ERK½ phosphorylation increased up to twofold after 90–180 min of sustained pressure exposure (Fig. 3.7a). In three of four independent experiments (samples 1–3), peak activation was observed at 180 min, while in one experiment (sample 4), the peak occurred earlier at 90 min (Fig. 3.7b). On average, ERK½ activation at 180 min was significantly higher (1.6-fold;

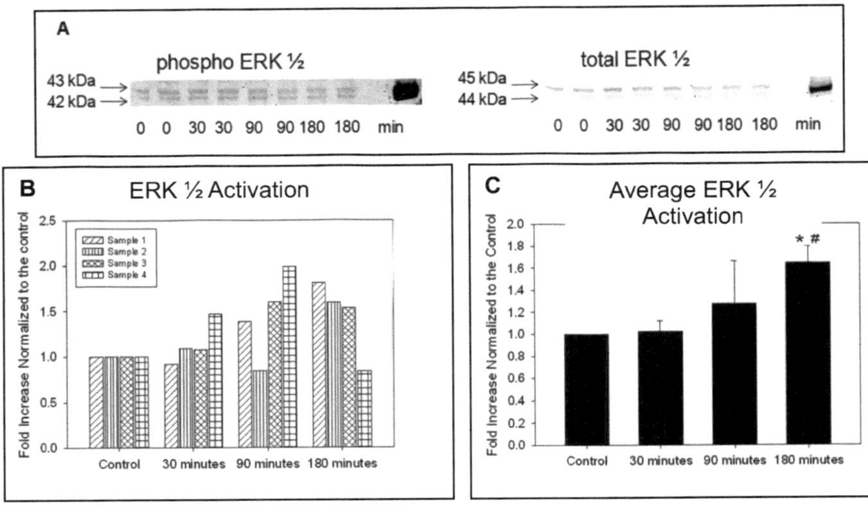

Fig. 3.7 ERK½ activation in BSMCs following sustained hydrostatic pressure. ERK½ phosphorylation was quantified by Western blot. (**a**) Representative chemiluminescence blot images. (**b**) Activation profile from four independent experiments; three experiments (Samples 1–3) peaked at 180 min, while one (Sample 4) peaked at 90 min. (**c**) Mean ERK½ phosphorylation levels in BSMCs exposed to sustained pressure compared to controls. Spot densitometry was used to calculate fold changes (phospho-ERK½ / total ERK½). Data are mean ± SD; ANOVA; *p < 0.05 vs. control; #p < 0.05 vs. 30-min group; n = 3 of 4 experiments. (Adapted with permission from *Urology*, Elsevier (2010))

$p < 0.05$, n = 3) compared to baseline (0 min) and 30 min (Fig. 3.7c). In contrast, expression of α-SMA and SM22 remained unchanged across all groups, regardless of pressure exposure or U0126 treatment, at 48 h (data not shown).

Cell proliferation was quantified by measuring adherent cell density (cells/mm^2) following pressure application. At 24 h, cell density in pressure-treated cultures was comparable to static controls (data not shown). However, after 48 h, pressure-treated BSMCs exhibited a significant increase in density ($p < 0.05$; n = 5) relative to controls (Fig. 3.8). Importantly, this proliferative effect was abolished when cells were pretreated with U0126, implicating MEK½ - ERK½ signaling in pressure-induced proliferation (Fig. 3.8).

3.3.3 Discussion

This study examined how sustained, physiological levels of hydrostatic pressure influence bladder smooth muscle cell (BSMC) phenotype in vitro. The experimental system, which used modified 50-mL conical tubes to apply a column of medium above the cell layer, was adapted from prior work (Haberstroh et al. 1999; Acevedo et al. 1993). This setup allowed the application of pressure while maintaining stable

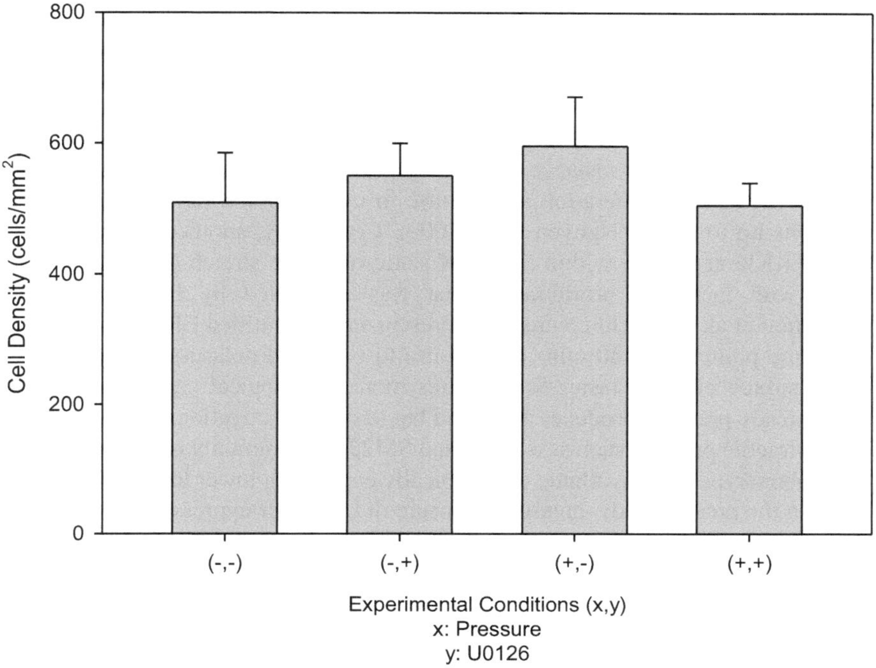

Fig. 3.8 BSMC proliferation after 48 h of sustained hydrostatic pressure. Cells were exposed to 7.5 cm H_2O for 48 h, fixed, and stained with DAPI to count nuclei. Cell density (cells/mm^2) was quantified using ImagePro software. Proliferation was significantly elevated in pressure-treated cells compared with controls. Data are mean ± SD; ANOVA and paired t-test; $^*p < 0.05$ vs. control; $n = 3–5$. (Adapted with permission from *Urology*, Elsevier (2010))

culture conditions, as confirmed by the absence of significant changes in pH, pO_2, and pCO_2 (Acevedo et al. 1993). Under these conditions, BSMCs exposed to 7.5 cmH_2O pressure for 4 h displayed a morphological shift from an elongated, spindle-shaped appearance toward a more rounded form. Such changes are consistent with a potential phenotypic transition from a contractile to a synthetic state. While this is the first report of BSMCs altering morphology in response to sustained pressure, our group has previously shown that BSMCs embedded in 3D collagen gels and subjected to 48 h of tensile loading developed higher aspect ratios—suggesting reinforcement of a contractile phenotype—relative to static controls (Roby et al. 2008). Taken together, these results highlight that the cellular phenotype of SMCs is highly dependent on the type of mechanical stimulus: stretch appears to promote a contractile state, while hydrostatic pressure promotes a synthetic-like phenotype.

Beyond morphology, exposure to sustained hydrostatic pressure also increased intracellular signaling and proliferation. Specifically, BSMCs demonstrated a two-fold elevation in ERK½ phosphorylation after 90–180 min of pressure exposure (Fig. 3.7), accompanied by a significant rise in cell density after 48 h (Fig. 3.8).

When cells were pretreated with the MEK½ inhibitor U0126, this pressure-induced proliferation was abolished, returning to levels comparable with no-pressure controls (Fig. 3.8). These findings strongly suggest that proliferation under pressure is mediated through the ERK½ pathway.

The involvement of ERK signaling in BSMC mechanotransduction, however, has been debated. For instance, one study reported that p38/SAPK, rather than ERK½, mediated proliferation in response to cyclic stretch (25% at 0.1 Hz, for durations up to 24 h) (Nguyen et al. 2000). Conversely, another group observed rapid ERK½ activation within 5 min of static or cyclic stretch (5–10%, 0.1 Hz), along with increased proliferation that was attenuated by ERK½ inhibition (Halachmi et al. 2006). In contrast, the present study identified ERK½ activation at later time points (90–180 min) in response to sustained pressure, suggesting that cyclic stretch elicits a faster and perhaps more pathological signaling response, while steady pressure produces a delayed but sustained activation.

Contractile proteins such as α-SMA and SM22 are commonly used as markers of SMC phenotype, with synthetic cells typically expressing lower levels of these proteins. In the present study, pressure exposure did not alter expression of either protein, even though proliferation increased. This observation aligns with previous work showing that stretch-induced expression of α-SMA and SM22 in vascular SMCs was abolished by inhibition of p38, but not ERK½ (Qu et al. 2007). These findings suggest that ERK½-mediated proliferation may occur independently of contractile protein regulation, which could explain why no changes in α-SMA or SM22 were observed here. Differences in tissue origin (vascular versus urinary tract) and experimental conditions further complicate direct comparisons, underscoring the need for additional studies to clarify the relationship between proliferation and phenotypic marker expression in bladder SMCs.

3.3.4 *Conclusions*

This study provides the first evidence that bladder SMCs respond to sustained physiological hydrostatic pressure with morphological changes suggestive of a contractile-to-synthetic phenotypic shift. The observed increase in cell proliferation was dependent on activation of the ERK½ signaling pathway, highlighting a possible mechanism by which pressure regulates BSMC behavior. These findings imply that exposure to low-level intravesical pressure oscillations, such as those occurring in developmentally compromised bladders, may contribute to detrusor muscle growth through ERK½-mediated proliferation.

3.4 The Effect of Hydrostatic Pressure on Neuronal Cell Morphology in Vitro

3.4.1 Introduction

Traumatic brain injury (TBI) is a neurological insult caused by external mechanical forces, with severities ranging from mild to life-threatening, and it is associated with both acute and long-term consequences (Caeyenberghs et al. 2014) TBIs occur in various contexts such as falls, vehicular accidents, assaults, sports impacts, and blast injuries, and represent nearly half of all injuries reported in combat zones. Following the initial insult, secondary damage often develops through biochemical cascades such as disruption of calcium homeostasis, increased oxidative stress, and lipid peroxidation (Xiong et al. 2009), or through biophysical changes such as intracranial hypertension and ischemia (Werner and Engelhard 2007). These events can lead to mitochondrial dysfunction, inflammation, and apoptosis in neurons (Xiong et al. 2009). In healthy adults, intracranial pressure (ICP) is typically between 5–15 mmHg (Rangel-Castillo et al. 2008). ICP elevations of 20–30 mmHg are associated with mild to moderate TBI, while pressures above 40 mmHg are linked to severe cases (Rangel-Castillo et al. 2008). Without prompt intervention, acute intracranial hypertension may be fatal (Hawryluk et al. 2022). Most studies of TBI have emphasized the primary mechanical insult, whereas the contribution of elevated ICP, particularly hydrostatic pressure (HP), to mechanotransduction processes remains underexplored.

Experimental TBI models have documented neuronal injury and neurite retraction caused by primary mechanical forces. For example, a 3D compression system using cortical neurons in collagen gels showed that mechanical impacts at different strain rates induced cytoskeletal deterioration, neurite thinning, blebbing, and eventual cell death within hours (Bar-Kochba et al. 2016). Similarly, micropatterned neuronal cultures on stretchable silicone membranes demonstrated that dynamic strain disrupted axonal microtubules, impaired axonal transport, and caused axonal swelling and degeneration (Tang et al. 2010).

Neurite retraction plays an important role in central nervous system injury, glaucoma, and neurodegenerative disorders. Previous work has implicated the Rho/ROCK pathway as a critical regulator of the actin cytoskeleton and neurite structure (Hirose et al. 1998; Sun et al. 2011; Monnier et al. 2003). Activation of Rho GTPases and their downstream effector Rho-kinase leads to disassembly of microtubules and intermediate filaments, driving neurite retraction (Hirose et al. 1998). Inhibitor studies using Y-27632, a selective blocker of p160ROCK, showed that suppressing this pathway prevented neurite retraction in N1E-115 neuroblastoma cells (Hirose et al. 1998; Monnier et al. 2003). Lysophosphatidic acid (LPA), another neurite retraction agonist, induces GSK-3 activation, and pharmacological inhibition of GSK-3 partially suppresses this effect (Sayas et al. 2006). Notably, RhoA appears to mediate GSK-3 activation in this process (Sayas et al. 2006). Although these studies reveal molecular players involved in neurite remodeling, how mechanical stimuli such as HP influence these pathways is not well understood.

Elevated HP is also a hallmark of other disorders such as glaucoma. Retinal ganglion cells (RGCs), CNS neurons whose somata reside in the retina and axons in the optic nerve (Almasieh et al. 2012), are a common glaucoma model. When differentiated RGC-5 cells were subjected to 100 mmHg HP in a live-cell chamber, intracellular Ca^{2+} spikes were observed, followed by caspase-3/7 activation. These biochemical changes corresponded with apoptosis-related morphology, including shrinkage and neurite retraction (Lee et al. 2010). Likewise, PC12 cells, derived from rat pheochromocytoma (Westerink and Ewing 2008), exposed to 15–70 mmHg HP displayed oxidative stress, loss of glutathione, mitochondrial depolarization, and apoptosis after prolonged exposure (Tök et al. 2014).

Despite these findings, the direct contribution of elevated ICP to neuronal structure and mechanotransduction remains unresolved. Morphological features such as neurite retraction and axonal degeneration have been consistently observed in TBI models, and these may underlie functional impairments and predispose patients to chronic neurodegeneration. This study therefore aimed to test the hypothesis that elevated HP directly alters neuronal morphology through a specific signaling pathway. Using Neuro2A cells and controlled pressurization, we examined neuronal responses to HP as a potential model of TBI-related pathophysiology.

3.4.2 *Summary of Experimental Results*

Mouse Neuro2A neuroblastoma cells were cultured in Dulbecco's Modified Eagle Medium (DMEM) with 10% fetal bovine serum (FBS) and 1% penicillin–streptomycin under standard conditions. For pressure studies, cells were plated at a density of 56,600 cells per dish on either tissue culture polystyrene (rigid substrate) or polyacrylamide–collagen coated glass coverslips (compliant substrate) prepared following published protocols (Fischer et al. 2012). Prior to experiments, cultures were serum-deprived for 24 h to stimulate neurite outgrowth (Fig. 3.9). A custom pressurization system was then used to apply 25 or 35 mmHg hydrostatic pressure (HP) for 30 min, 1 h, 2 h, or 4 h. Controls were handled identically but maintained under atmospheric conditions. Phase-contrast images of the same cells were collected before and immediately after pressurization (Fig. 3.10).

Neurite lengths were quantified using Fiji with the Simple Neurite Tracer plugin (Longair et al. 2011). For each cell, neurite number and combined length were measured before and after treatment. Normalized length change was calculated as:

$$\Delta L = \frac{\sum_{n} \frac{l}{N_0}}{\sum_{n_0} \frac{l_0}{N_0}}$$

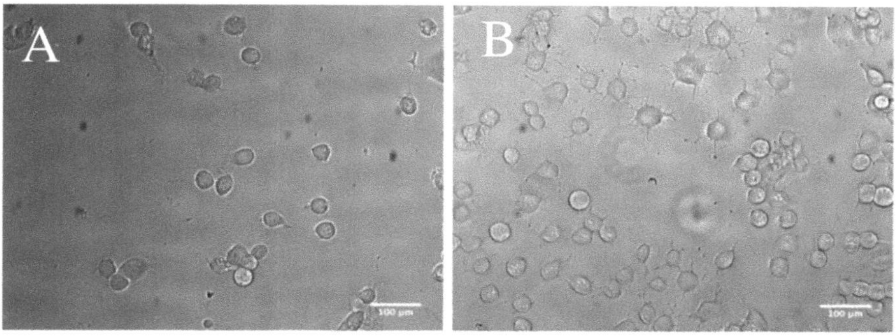

Fig. 3.9 Phase contrast micrographs of Neuro2A cells cultured in (**a**) complete growth medium (DMEM supplemented with 10% FBS and 1% Penicillin-Streptomycin) for 24 h, and (**b**) serum-free DMEM for 24 h. Scale bar = 100 μm. (Adapted from MS Thesis, Clemson University (2017))

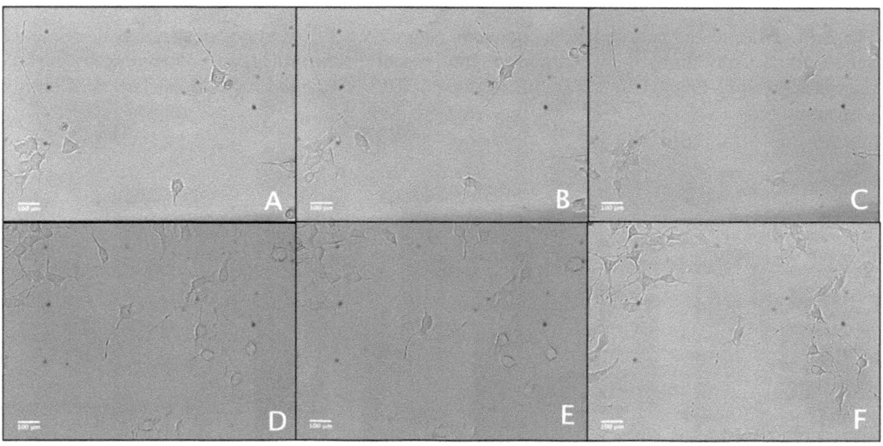

Fig. 3.10 Phase contrast images of Neuro2A cells maintained in serum-free DMEM. (**a**, **d**) baseline (time = 0); (**b**, **c**) incubator control with no pressure for 15 and 30 min, respectively; (**e**, **f**) cells exposed to 35 mmHg hydrostatic pressure for 15 and 30 min, respectively. Scale bar = 100 μm. (Adapted from MS Thesis, Clemson University (2017))

where l_0 is the combined length before, N_0 is the number of neurites before, and l is the combined length after. A ΔL of 1 indicates no change, values >1 indicate growth, and values <1 indicate retraction. Measurements (~30 cells per dish) revealed that both pressure groups trended toward reduced neurite length compared to controls. However, statistically significant ($p < 0.05$) reductions were detected only in cells exposed to 35 mmHg for 30 or 60 min (Fig. 3.11). Time-course analysis further confirmed significant shortening at 30 min, 60 min, and 4 h of 35 mmHg exposure (Fig. 3.12), suggesting a time dependent effect.

To test a brain like microenvironment, Neuro2A cells were seeded on soft collagen coated polyacrylamide substrates. Compared to polystyrene, cells appeared

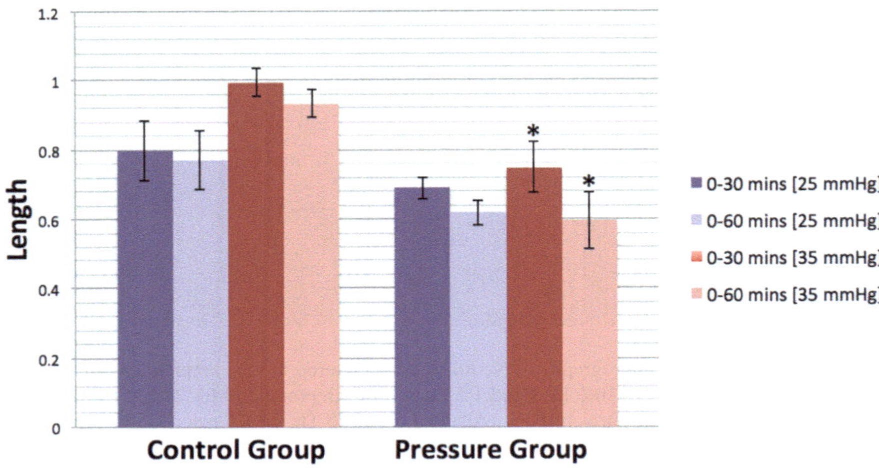

Fig. 3.11 Normalized neurite length change in Neuro2A cells over two consecutive 30-min intervals. Comparisons shown between control (no pressure), 25 mmHg, and 35 mmHg hydrostatic pressure groups. Error bars = ± S.E.M. *$p < 0.05$. (Adapted from MS Thesis, Clemson University (2017))

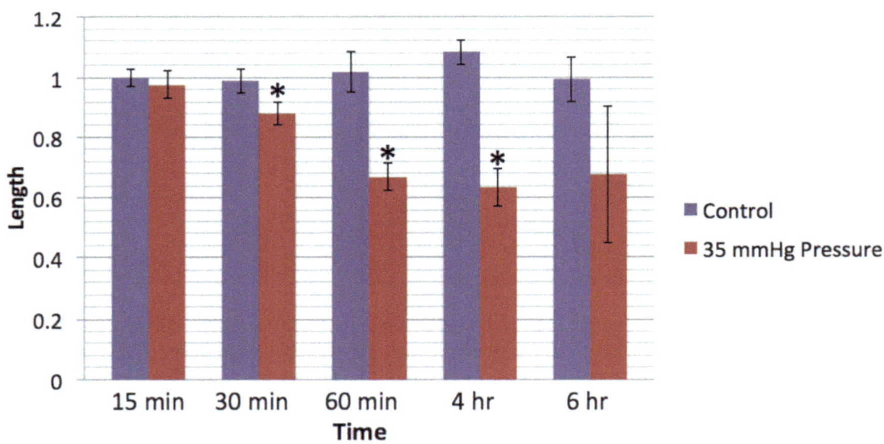

Fig. 3.12 Time-dependent normalized neurite length changes in Neuro2A cells subjected to 35 mmHg hydrostatic pressure compared with control (no pressure). Error bars = ± S.E.M. *$p < 0.05$. (Adapted from MS Thesis, Clemson University (2017))

more rounded with fewer extensions (Fig. 3.13). After 1 h of 35 mmHg HP, neurite length was significantly reduced on the soft substrate as well ($p < 0.05$) (Fig. 3.14).

Finally, mechanistic assays were conducted to evaluate involvement of the Rho/ROCK pathway. Neuro2A cells were pretreated with the ROCK inhibitor Y 27632 (1 μM, 30 min) prior to 35 mmHg HP (1 h). LPA (0.3 mM), a known inducer of neurite retraction, served as a positive control. Quantification revealed that while

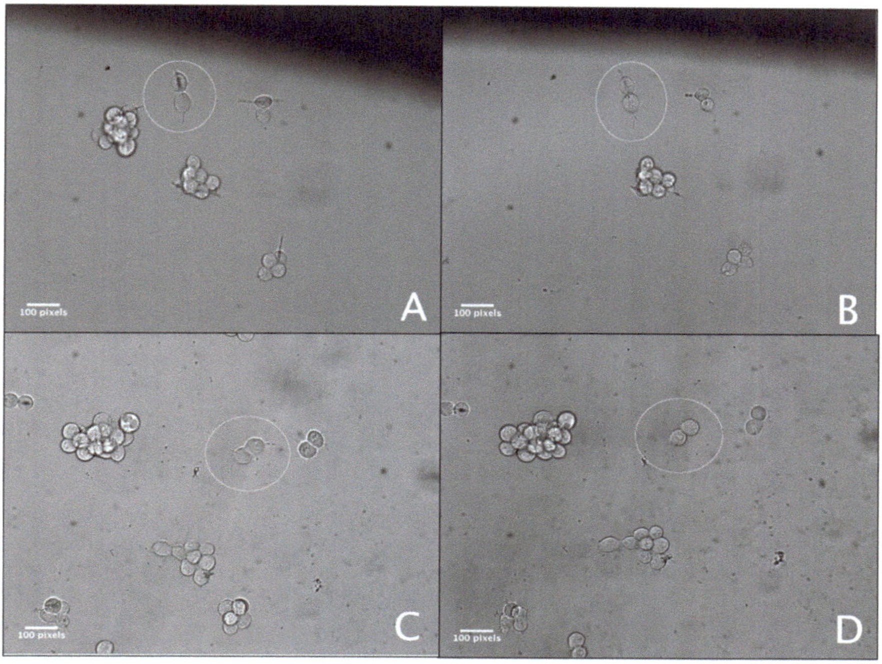

Fig. 3.13 Phase contrast images of Neuro2A cells cultured on a soft polyacrylamide + collagen I substrate. (**a**, **c**) baseline (time = 0); (**b**) incubator control after 1 h (no pressure); (**d**) cells exposed to 35 mmHg hydrostatic pressure for 1 h. Scale bar = 100 μm. (Adapted from MS Thesis, Clemson University (2017))

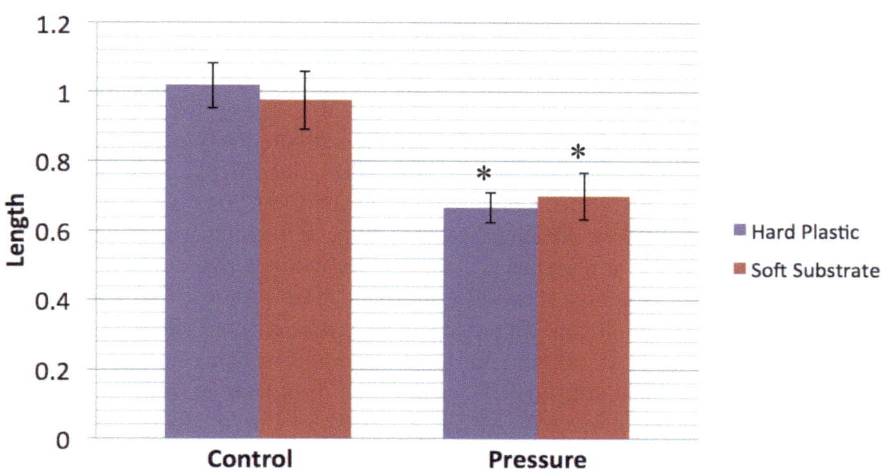

Fig. 3.14 Normalized neurite length changes in Neuro2A cells cultured on hard versus soft substrates over 1 h. Comparison between control (no pressure) and 35 mmHg pressure-exposed groups. Error bars = ± S.E.M. $*p < 0.05$. (Adapted from MS Thesis, Clemson University (2017))

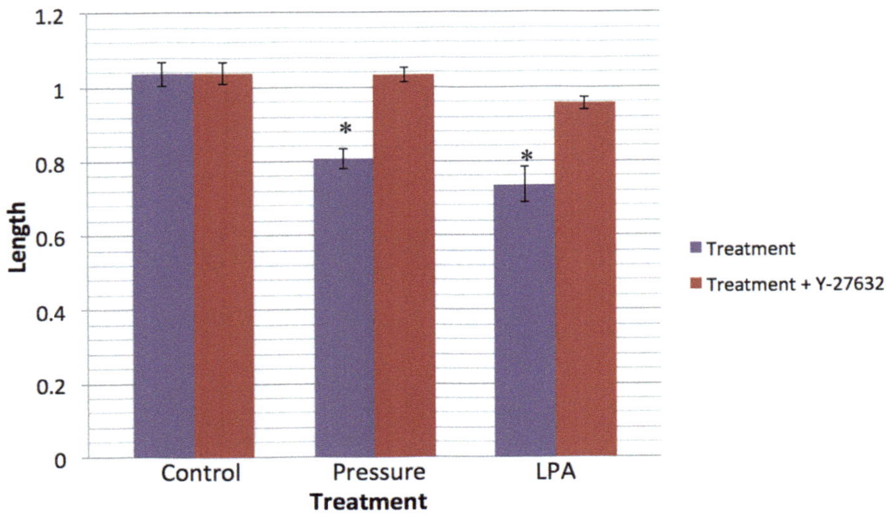

Fig. 3.15 Normalized neurite length changes in Neuro2A cells across experimental groups over 1 h. Groups include: control (no pressure), control + Y-27632, 35 mmHg pressure, 35 mmHg pressure + Y-27632, LPA treatment, and LPA + Y-27632. Error bars = ± S.E.M. *$p < 0.05$. (Adapted from MS Thesis, Clemson University (2017))

untreated and LPA treated groups underwent significant retraction, cells pretreated with Y 27,632 showed minimal change. Moreover, Y 27632 also blocked LPA induced retraction (Fig. 3.15). These findings suggest that pressure induced neurite retraction is mediated through the Rho/ROCK pathway.

3.4.3 Discussion

Although many investigations have examined the pathological mechanisms underlying traumatic brain injury (TBI) (Xiong et al. 2009; Bar-Kochba et al. 2016; Tang et al. 2010; Wu et al. 2012; Prado et al. 2005; Gao and Chen 2011; DiLeonardi et al. 2009; Casella et al. 2014), none have directly evaluated the impact of elevated intracranial pressure (ICP) on neuronal mechanobiology. Since there are currently no therapeutic options targeting secondary injury in TBI—some of which may be driven by elevated ICP—identifying mechanistic pathways could help guide future interventions (Xiong et al. 2009). Clarifying these mechanisms may also support the development of standardized ICP management protocols in TBI, an area that remains debated among clinicians (Alali et al. 2013). In vitro assessments of neuronal cell health frequently focus on morphological characteristics, such as neurite growth and retraction, or on apoptotic outcomes, as indicators of cellular viability (Hirose et al. 1998; Monnier et al. 2003; Sayas et al. 2006). The present study used neurite length as a quantitative measure to evaluate how elevated hydrostatic pressure (HP), modeling ICP, influences Neuro2A cells. We hypothesized that elevated HP would cause neurite retraction in this in vitro model.

Neuroblastoma cell lines are commonly employed to study neuronal behavior (Seeds et al. 1970; Schubert et al. 1974). Several approaches are available to induce differentiation and neurite outgrowth, with serum deprivation being the most straightforward (Seeds et al. 1970; Schubert et al. 1974; Schubert et al. 1971; Gurwitz and Cunningham 1988). Removal of serum supports protein synthesis necessary for microtubule assembly during neurite extension (Seeds et al. 1970) and may also promote stronger attachment to the culture surface, thereby enabling neurite elongation (Schubert et al. 1974; Schubert et al. 1971). In this study, Neuro2A cells were grown in complete medium (DMEM with 10% FBS and 1% penicillin–streptomycin) before being switched to serum-free DMEM for 24 h, which successfully induced neurite outgrowth. Differentiation was essential to obtain a neuronal-like phenotype with greater clinical relevance.

The effects of HP on cell morphology were assessed primarily in standard polystyrene dishes, with additional tests performed on a compliant substrate composed of polyacrylamide with a type I collagen coating to approximate brain tissue stiffness. Differentiated Neuro2A cells were subjected to 25 or 35 mmHg of HP for 15 min, 30 min, 60 min, 4 h, or 6 h. These pressure magnitudes were chosen to correspond to ICP values reported for moderate and severe TBI (Hawryluk et al. 2022). Phase-contrast images, analyzed using Fiji with the Simple Neurite Tracer plugin (Longair et al. 2011), showed evidence of neurite retraction and extension in both control and pressure-exposed groups. Some of these changes likely reflected natural cell migration or transient responses to handling, such as removal from the incubator. Thus, comparisons between experimental and control groups were used to determine the effects specifically attributable to pressure.

Results indicated that cells exposed to 35 mmHg HP exhibited greater neurite shortening than those exposed to 25 mmHg, particularly at 30 and 60 min (Fig. 3.11). This trend suggests that neurite retraction begins at moderate ICP levels (~25 mmHg) and becomes more pronounced with increasing pressure. Time-course analysis further demonstrated that significant retraction occurred after 30 min, 60 min, and 4 h of 35 mmHg exposure. Interestingly, at 6 h, neurite shortening appeared less pronounced than at 4 h. While neurite regrowth may have begun, this observation was more likely due to variability introduced by increased cell migration over the longer culture period.

These findings imply that neurons within brain tissue exposed to severe TBI-associated ICP may undergo neurite retraction. Such retraction could disrupt synaptic integrity and neuronal communication (Gao and Chen 2011), potentially contributing to long-term outcomes observed in TBI patients, including neurodegenerative conditions (Chen et al. 2007; Fleminger et al. 2003; Jafari et al. 2013; Mortimer et al. 1991) and cognitive impairments (Caeyenberghs et al. 2014), which are also linked to neurite retraction (Radio et al. 2008; Sheinerman and Umansky 2013). Our results further support the clinical importance of rapidly lowering ICP—ideally within 30 min of elevation to ~35 mmHg—to prevent neuronal damage.

While neurite retraction was the focus here, additional pathological processes may also occur under these conditions.

Previous work has highlighted the influence of substrate stiffness on cell morphology and differentiation (Engler et al. 2006; Yeung et al. 2005). Given that brain tissue has a shear modulus of ~0.49 kPa (Rashid et al. 2013)—orders of magnitude lower than polystyrene dishes (~1 GPa) (Syed et al. 2015)—it was important to verify findings on a softer substrate. Using polyacrylamide–collagen-coated coverslips, we confirmed that Neuro2A cells adhered successfully via immobilized type I collagen. Cells on this soft substrate displayed less neurite extension compared with those on polystyrene (Fig. 3.13). Nonetheless, significant neurite retraction was observed after 1 h of exposure to 35 mmHg HP (Fig. 3.14), validating the rigid substrate experiments as representative in vitro models of elevated ICP.

Prior studies have shown that the Rho/ROCK pathway mediates neurite retraction induced by stimuli such as lysophosphatidic acid (LPA) (Hirose et al. 1998; Monnier et al. 2003), but no research had addressed how HP engages this pathway. In our experiments, pretreatment with the ROCK inhibitor Y-27632 significantly reduced neurite retraction after 1 h at 35 mmHg compared with untreated cells ($p < 0.05$). As expected, LPA induced neurite retraction, but this effect was also blocked by Y-27632 pretreatment. These observations suggest that HP-induced neurite retraction is mediated by the same Rho/ROCK signaling pathway activated by LPA.

While this in vitro system allowed precise control of HP, several limitations remain. First, neurite measurements did not distinguish between axons and dendrites, which were collectively analyzed for simplicity. Second, the culture conditions used—95% air, 5% CO_2, 37 °C—mimic hyperventilation, whereas cortical neurons in vivo are normally exposed to lower oxygen levels (Ndubuizu and LaManna 2007). Future studies should adjust oxygen concentrations to more closely reflect physiological conditions. Additionally, pressure experiments were conducted in serum-free medium to induce differentiation, but in vivo neurons under elevated ICP would still receive nutrients from surrounding brain tissue. Using nutrient-rich conditions may produce different outcomes.

In summary, our findings demonstrate that exposure to 35 mmHg HP for at least 30 min induces neurite retraction in Neuro2A cells, and that this effect is mediated through the Rho/ROCK pathway. These results provide new insights into how elevated ICP may contribute to neuronal damage in TBI. Future work should assess whether HP also promotes apoptosis, identify specific cytoskeletal elements involved in neurite responses, and incorporate co-culture models with neurons, astrocytes, oligodendrocytes, and microglia to better replicate the brain microenvironment.

3.5 Hydrostatic Pressure Mechanotransduction in Bladder Urothelial Cells and the Role of Ion Channels

3.5.1 *Introduction*

The bladder urothelium, which lines the inner surface of the urinary bladder, is now recognized as more than just a passive barrier. Increasing evidence shows that it actively participates in sensing mechanical forces, releasing signaling molecules, and communicating with underlying smooth muscle and afferent nerves (Lewis 2000; Wang et al. 2003; Apodaca 2004; Gevaert et al. 2007; Lindstroêm and Mazieres 1991; Birder et al. 2002; Ferguson et al. 1997). One key mediator in this process is adenosine triphosphate (ATP), which can be released by urothelial cells (UCs) in response to mechanical stretch, shear stress, or elevated pressure within the bladder (Birder et al. 2002; Ferguson et al. 1997; Matsumoto-Miyai et al. 2009; Vlaskovska et al. 2001; Yu et al. 2009; Birder et al. 2003). Once released, ATP acts in an autocrine and paracrine fashion, activating purinergic receptors that contribute to bladder sensation and reflex activity. Several studies have implicated mechanosensitive ion channels in regulating this ATP release. Transient receptor potential (TRP) channels, such as TRPV1 (Birder et al. 2002) and TRPV4 (Mochizuki et al. 2009), as well as epithelial sodium channels (ENaC) (Du et al. 2007), are thought to play important roles in mediating cellular responses to changes in pressure and stretch. However, the specific contribution of these channels to pressure-induced ATP release from UCs remains incompletely understood. To address this gap, we designed experiments to examine the effect of hydrostatic pressure on ATP release from rat urothelial cells. In particular, we investigated whether pressure-induced ATP release depends on extracellular calcium, and how selective inhibition of mechanosensitive channels—including ENaC and stretch-activated channels—affects this process. We also examined the effect of TRPV1 and TRPV4 agonists in stimulating ATP release in the absence of pressure.

The findings presented in this section are intended to clarify the mechanistic pathways by which UCs sense and respond to elevated hydrostatic pressure. Such insights may have implications for understanding bladder pathophysiology, including overactive bladder and bladder outlet obstruction, conditions in which altered mechanotransduction and ATP signaling are suspected to play major roles.

3.5.2 *Summary of Experimental Results*

Primary rat urothelial cells (UCs) were harvested according to a published method (Kurzrock et al. 2005) and characterized. UCs were exposed to defined levels of hydrostatic pressure using a custom-built chamber (Stover and Nagatomi 2007). Pressure exposures ranged from 5 to 20 cmH_2O and were applied for durations of 5, 15, or 30 min. Culture supernatants were collected immediately following exposure and analyzed for ATP concentration. Controls consisted of UCs maintained under atmospheric pressure.

Application of hydrostatic pressure led to a time-and pressure-dependent increase in ATP release. Significant elevations in extracellular ATP were observed after 5 min of exposure at 10 and 15 cmH2O, and after 15 min at 15 cmH2O. ATP levels were normalized to baseline and expressed as fold change. By 30 min, ATP release generally returned toward control levels, suggesting a transient adaptive response (Fig. 3.16).

To test whether calcium influx was necessary for this response, cells were pretreated with the calcium chelator BAPTA (1 mM). In these conditions, pressure-induced ATP release was abolished, indicating that extracellular calcium entry is required (Fig. 3.17). Further mechanistic insight was obtained using pharmacological inhibitors. Pretreatment with gadolinium chloride (50 μM), a broad blocker of stretch-activated channels, significantly reduced pressure-induced ATP release (Fig. 3.18). Similar inhibitory effects were observed with ruthenium red (1 μM), a nonspecific TRP channel blocker, and amiloride (1 mM), an ENaC inhibitor (Fig. 3.18). By contrast, nifedipine (10 μM), an L-type voltage-gated calcium channel blocker, had no significant effect (Fig. 3.18). These results suggest that ATP release in response to hydrostatic pressure occurs predominantly through mechanosensitive channels rather than voltage-gated calcium channels. Finally, to verify

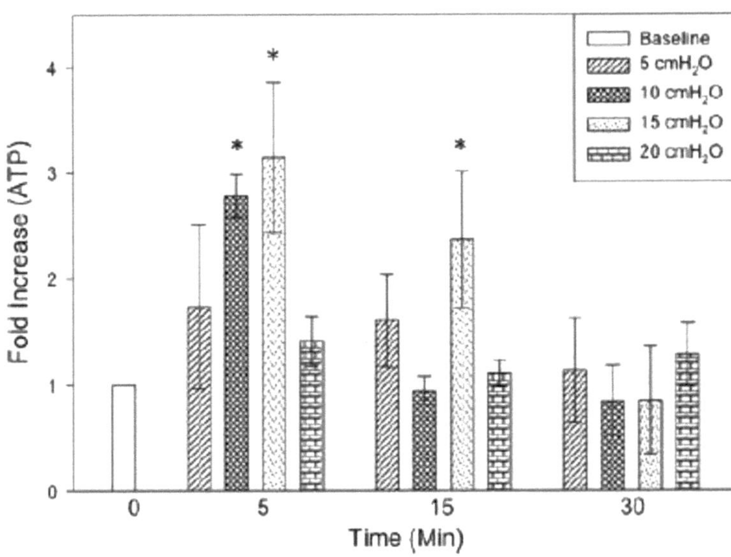

Fig. 3.16 Time-dependent ATP release from urothelial cells exposed to hydrostatic pressure. Supernatant media were collected following stimulation with pressures ranging from 5–20 cmH_2O for 5, 15, and 30 min. Cells exposed for 5 min (10 and 15 cmH_2O) and for 15 min (15 cmH_2O) showed significantly higher ATP release compared to controls. ATP values are expressed as fold-change relative to baseline. Data are presented as mean ± SEM; *$p < 0.05$, ANOVA with Tukey's post-hoc test, n = 3 experiments. (Adapted with permission from Ann Biomed Eng, Springer (2011))

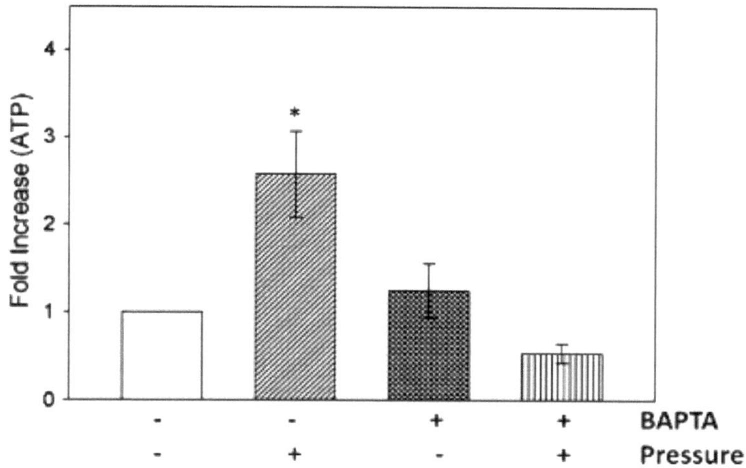

Fig. 3.17 Effect of calcium chelation on pressure-induced ATP release. Rat urothelial cells were subjected to 10 cmH_2O hydrostatic pressure for 5 min, with or without pretreatment using BAPTA (1 mM). Pressure alone increased ATP release, whereas cells treated with BAPTA did not exhibit this response. Data are shown as mean ± SEM; $^*p < 0.05$, ANOVA with Tukey's post-hoc test, n = 3 experiments. (Adapted with permission from Ann Biomed Eng, Springer (2011))

TRP channel functionality in UCs, agonists were applied in the absence of pressure. Capsaicin (10 μM, TRPV1 agonist) and 4α-PDD (10 μM, TRPV4 agonist) each triggered a significant increase in ATP release compared to control (Fig. 3.18), supporting the presence of functional TRPV1 and TRPV4 channels in urothelial cells.

3.5.3 *Discussion*

This study provides new insight into how urothelial cells (UCs) sense and respond to hydrostatic pressure, a stimulus relevant to bladder filling and pathological states such as bladder outlet obstruction. The data demonstrate that exposure of rat UCs to modest elevations in hydrostatic pressure results in a rapid but transient increase in ATP release, with the strongest responses observed at 10–15 cmH_2O after 5–15 min (Fig. 3.16). These findings are consistent with the concept that the urothelium functions as a mechanosensory tissue (Lewis 2000; Wang et al. 2003; Apodaca 2004; Gevaert et al. 2007; Lindstroêm and Mazieres 1991; Birder et al. 2002; Ferguson et al. 1997; Matsumoto-Miyai et al. 2009; Vlaskovska et al. 2001; Yu et al. 2009; Birder et al. 2003; Mochizuki et al. 2009; Du et al. 2007), translating physical pressure into chemical signals that can influence underlying nerves and smooth muscle. The experiments further identified extracellular calcium as a critical requirement for pressure-induced ATP release. Chelation with BAPTA abolished the response, suggesting that calcium influx through mechanosensitive pathways is necessary for

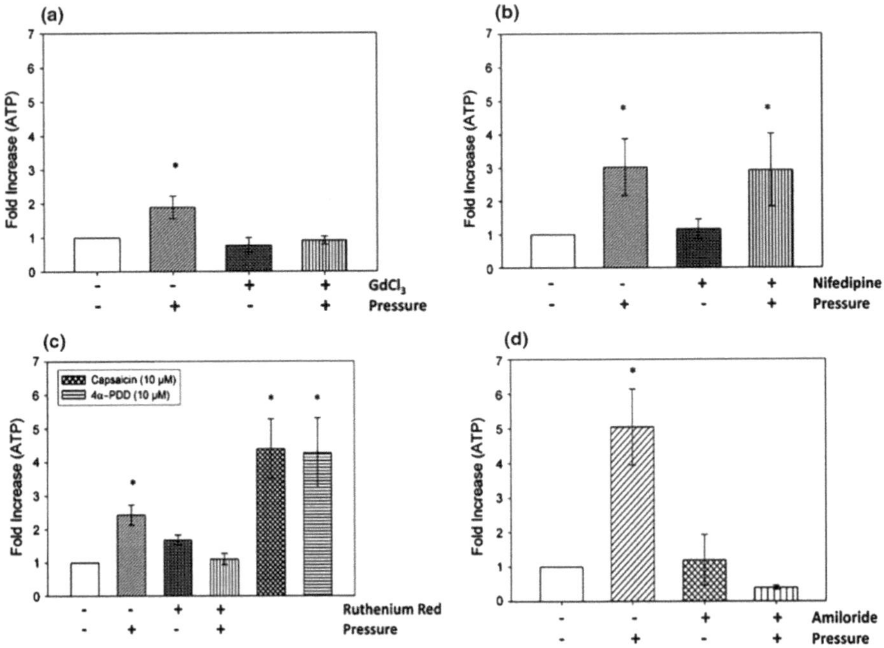

Fig. 3.18 ATP release in response to hydrostatic pressure in the presence of mechanosensitive ion channel blockers. Rat urothelial cells were exposed to 10 cmH_2O hydrostatic pressure for 5 min, with or without pretreatment with GdCl3 (Frame A, 50 μM), nifedipine (Frame B, 10 μM), ruthenium red (Frame C, 1 μM), or amiloride (Frame D, 1 mM). Gadolinium chloride, ruthenium red, and amiloride significantly reduced ATP release compared to pressure alone, while nifedipine did not. In separate experiments, cells were stimulated with TRPV1 agonist capsaicin (10 μM) and TRPV4 agonist 4α-PDD (10 μM) without pressure (Frame C), both of which significantly enhanced ATP release. Data are mean ± SEM; *$p < 0.05$, ANOVA with Tukey's post-hoc test, $n = 3$ experiments. (Adapted with permission from Ann Biomed Eng, Springer (2011))

transducing hydrostatic force into ATP release (Fig. 3.18). This aligns with prior evidence that calcium entry is central to mechanotransduction in epithelial systems. Pharmacological inhibition studies clarified the ion channels involved. The broad stretch-activated channel blocker gadolinium chloride, along with ruthenium red and amiloride, each suppressed pressure-induced ATP release, indicating the participation of multiple channel families (Fig. 3.18). In particular, these results implicate members of the TRP family and ENaC in mechanosensitive signaling within UCs. By contrast, nifedipine, an L-type calcium channel antagonist, had no significant effect, suggesting that voltage-gated calcium entry is not the principal pathway in this setting (Fig. 3.18).

Additional support for TRP channel involvement was provided by agonist experiments. Application of capsaicin (TRPV1 agonist) and 4α-PDD (TRPV4 agonist) triggered ATP release in the absence of pressure, confirming that UCs express functional TRPV1 and TRPV4 channels capable of stimulating purinergic signaling. These findings indicate that TRP channels, together with ENaC, play a substantial role in transducing mechanical pressure into ATP release.

Collectively, these results support a model in which mechanosensitive ion channels mediate calcium-dependent ATP release from urothelial cells under hydrostatic load. The transient nature of the response suggests an adaptive mechanism that may prevent continuous ATP release during sustained bladder filling, instead producing bursts of purinergic signaling at pressure thresholds relevant to bladder sensation. From a physiological standpoint, this mechanism may contribute to the initiation of bladder afferent activity and the regulation of voiding reflexes. Pathologically, exaggerated or dysregulated ATP release could underlie sensory abnormalities observed in conditions such as overactive bladder or bladder outlet obstruction.

Future work should extend these findings by testing primary human UCs and by evaluating the interaction between urothelial cells and afferent nerves under controlled pressure environments. Investigating whether chronic exposure to elevated hydrostatic pressure alters channel expression or ATP release dynamics will also be important for understanding disease progression.

3.6 Elevated Hydrostatic Pressure Stimulates ATP Release which Mediates Activation of the NLRP3 Inflammasome Via P2X$_4$ in Rat Urothelial Cells

3.6.1 Introduction

Bladder urothelial cells are naturally exposed to hydrostatic pressure during normal filling and voiding, and these forces are considered essential for bladder development and function. However, under pathological conditions such as partial bladder outlet obstruction (pBOO), commonly due to prostate enlargement in aging men or pelvic organ prolapse in women, intravesical pressures can rise abnormally. pBOO is frequently associated with bladder wall inflammation and fibrosis (Metcalfe et al. 2010). Animal studies have further shown that pBOO activates the inflammasome, resulting in increased expression of pro-inflammatory cytokines such as interleukin-1β (IL-1β) (Metcalfe et al. 2010; Oka et al. 2009; Hughes et al. 2016). Despite these associations, the signaling cascade that connects mechanical stress from elevated intravesical pressure to inflammasome activation remains poorly defined.

Clinical and animal study data highlight the altered pressure environment in pBOO. Specifically, detrusor pressures during storage ranged from 3–16 cmH_2O in unobstructed bladders versus 3–25 cmH_2O in obstructed ones, while voiding pressures increased from 17–48 cmH_2O to 35–88 cmH_2O (Han et al. 2015; Pandita et al. 2000; Hughes et al. 2014). Parallel in vitro experiments (see Sect. 3.5) revealed that primary urothelial cells from mice and rats release ATP in response to hydrostatic pressure elevations up to 20 cmH_2O (Olsen et al. 2011), mechanical distension (~127%) (Mochizuki et al. 2009), or combined pressure and stretch in Ussing chamber studies (Birder et al. 2002). These findings support the idea that multiple forms of mechanical stress present in the obstructed bladder can trigger urothelial ATP release, which may in turn amplify downstream signaling.

ATP is a multifunctional signaling molecule that regulates a broad range of cellular responses, including inflammasome activation and tissue fibrosis through pro-inflammatory cytokines (Riteau et al. 2010; Cauwels et al. 2014). Hughes et al. demonstrated that the NLRP3 inflammasome plays a central role in inflammation and hypertrophy in a rat model of pBOO (Hughes et al. 2016). It is now well established that NLRP3 can be activated by pathogen-associated molecular patterns (PAMPs) such as LPS and flagellin (Lamkanfi et al. 2009), as well as DAMPs including ATP (Roose-Girma et al. 2006). Once activated, NLRP3 facilitates the cleavage of pro-caspase-1 into its active form, which in turn processes pro-IL-1β into mature IL-1β (Schroder and Tschopp 2010). Based on these findings, we hypothesized that the elevated intravesical pressures associated with pBOO promote ATP release from urothelial cells, which then activates caspase-1 via purinergic receptors, ultimately driving urothelial inflammation. To test this hypothesis, we investigated pressure-induced ATP signaling and NLRP3-mediated caspase-1 activation using a cultured rat urothelial cell line exposed to controlled hydrostatic pressure in vitro.

3.6.2 Summary of Experimental Results

To investigate the role of purinergic signaling in pressure-induced responses of bladder urothelial cells, we employed both a rat model of pBOO, as described previously (Hughes et al. 2016), and an in vitro culture model. MYP3 cells, an immortalized non-tumorigenic rat urothelial cell line (Kawamata et al. 1993), were grown in F-12 K medium supplemented with 10% fetal bovine serum until ~90% confluence, followed by an 18-h incubation in serum-free medium to achieve quiescence. Using a custom-built pressure chamber (Stover and Nagatomi 2007), cells were subjected either to 15 cmH_2O for 60 min to simulate pathological storage pressures or to 40 cmH_2O for 1 min to mimic obstructed voiding pressures. Cells maintained under atmospheric conditions served as controls, while exposure to hypotonic solution (240 mOsm for 60 min) was used as a positive control. Following treatment, supernatant was collected to quantify extracellular ATP release, and lysates were analyzed for intracellular caspase-1 activity. Additional experiments were conducted with pharmacological inhibitors to elucidate mechanisms of ATP release and caspase-1 activation (Table 3.1).

MYP3 cells exposed to elevated hydrostatic pressure—either 15 cmH_2O for 60 min or 40 cmH_2O for 1 min—released significantly more ATP compared with controls (Fig. 3.19a). Hypotonic swelling also enhanced ATP release (Fig. 3.19a). Both pressure and hypotonic stress led to increased intracellular caspase-1 activity (Fig. 3.19b). A dose–response experiment confirmed that exogenous ATP (ranging from 2.5 to 500 μM) caused a concentration-dependent rise in caspase-1 activity, suggesting that ATP release is directly linked to inflammasome activation (Fig. 3.20a).

Table 3.1 Pharmacological agents used to probe ATP release and caspase-1 signaling

Drug	Concentration	Mechanism of Action
5-BDBD	20 μM	P2X4 antagonist
A-438079	10 μM	P2X7 antagonist
PPADS	25 μM	Non-specific purinergic antagonist
Apyrase	1 U/mL	Extracellular ATP hydrolysis (ATPase)
Brefeldin A	10 ng/mL	Inhibitor of vesicular exocytosis
Probenecid	100 μM	Ion channel blocker

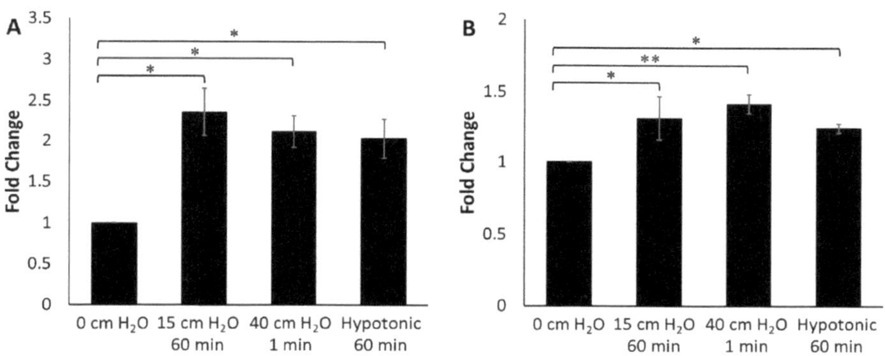

Fig. 3.19 Hydrostatic pressure elevates extracellular ATP and caspase-1 activity. MYP3 cells were subjected to hydrostatic pressures of 15 cmH_2O for 60 min or 40 cmH_2O for 1 min to model pathological storage and voiding conditions, respectively. After 60 min of incubation, culture media were collected for quantification of extracellular ATP (A), and cell lysates were analyzed for caspase-1 activity (B). Cells maintained under atmospheric pressure (0 cmH_2O, 60 min) served as baseline controls, while cells treated with hypotonic solution (240 mOsm, 60 min) served as positive controls. Results are expressed as fold change relative to the 0 cmH_2O control. Data represent mean ± SEM, with group comparisons evaluated using one-way ANOVA followed by Tukey's post hoc test. $^*P < 0.05$, $^{**}P < 0.01$. Sample size: $n = 14$, except $n = 8$ for the 15 cmH_2O group. (Adapted with permission from *Int Urol Nephrol* (Springer, 2018))

To test whether vesicular or transporter-mediated release mechanisms contribute to this process, cells were pretreated with a cocktail of Brefeldin A, Probenecid, and Apyrase. This combination significantly reduced extracellular ATP concentrations under both control and pressure conditions (40 cmH_2O for 1 min) (Fig. 3.20b). Importantly, caspase-1 activity did not increase in cocktail-treated cells exposed to pressure (Fig. 3.20c), indicating that ATP release and its persistence in the extracellular space are required for downstream activation.

Immunostaining revealed expression of both P2X7 and P2X4 receptors in rat bladder tissue and in cultured MYP3 cells, with localization primarily at the plasma membrane (Fig. 3.21). Pharmacological inhibition experiments further clarified

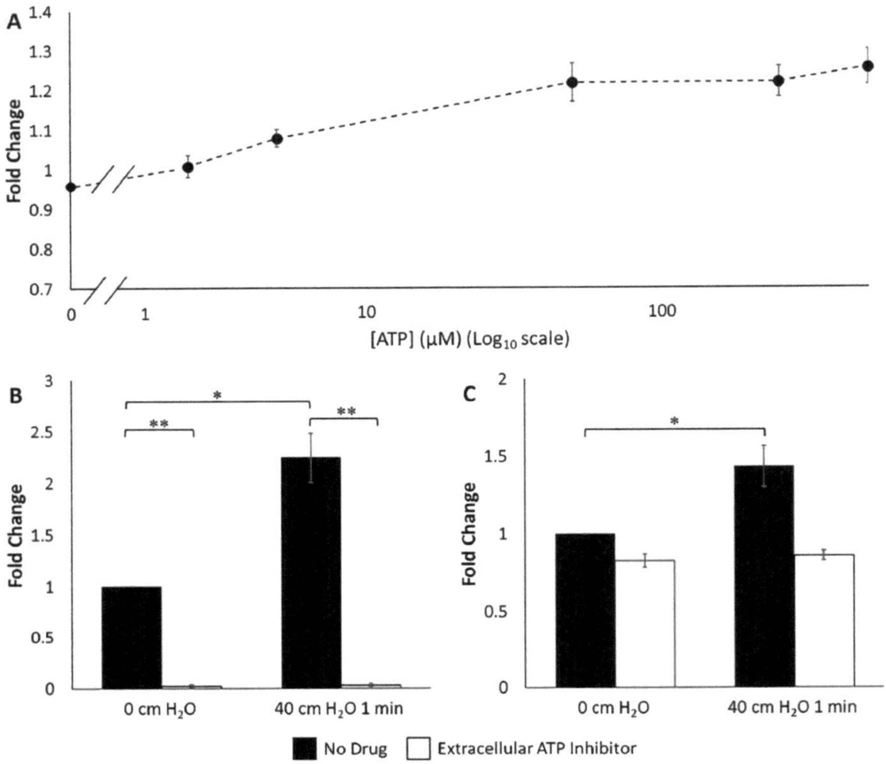

Fig. 3.20 Extracellular ATP removal and inhibition of ATP release prevent pressure-induced caspase-1 activation. (**a**) MYP3 cells exposed to increasing concentrations of ATP disodium salt for 60 min exhibited dose-dependent caspase-1 activity. (**b**, **c**) To assess the mechanism of ATP release, cells were divided into two groups: untreated or pretreated with a cocktail of inhibitors (1 U/mL apyrase, 100 µM probenecid, 10 µg/mL brefeldin A). Cells were further incubated with A-438079 (10 µM), 5-BDBD (20 µM), or PPADS (25 µM) prior to pressure challenge (40 cmH_2O, 1 min). Following exposure, extracellular ATP levels (**b**) and intracellular caspase-1 activity (C) were determined. Data are normalized to the 0 cmH_2O, no-drug control and presented as fold change (mean ± SEM). Statistical comparisons were made using one-way ANOVA with Tukey's test. *$P < 0.05$, $n = 4$. (Adapted with permission from *Int Urol Nephrol* (Springer, 2018))

their roles. Cells treated with PPADS (non-specific antagonist) or A-438079 (P2X7 antagonist) exhibited reduced ATP release and caspase-1 activity under pressure (Fig. 3.22). In contrast, 5-BDBD (P2X4 antagonist) did not affect ATP release but significantly decreased caspase-1 activity (Fig. 3.22). These findings suggest a two-step mechanism: P2X7 contributes to ATP-dependent ATP release, whereas P2X4 mediates ATP-driven inflammasome activation downstream.

Further experiments with exogenous ATP (500 µM) reinforced this conclusion. Caspase-1 activity was not altered by P2X7 blockade but was significantly reduced by either non-specific inhibition (PPADS) or selective P2X4 inhibition (5-BDBD)

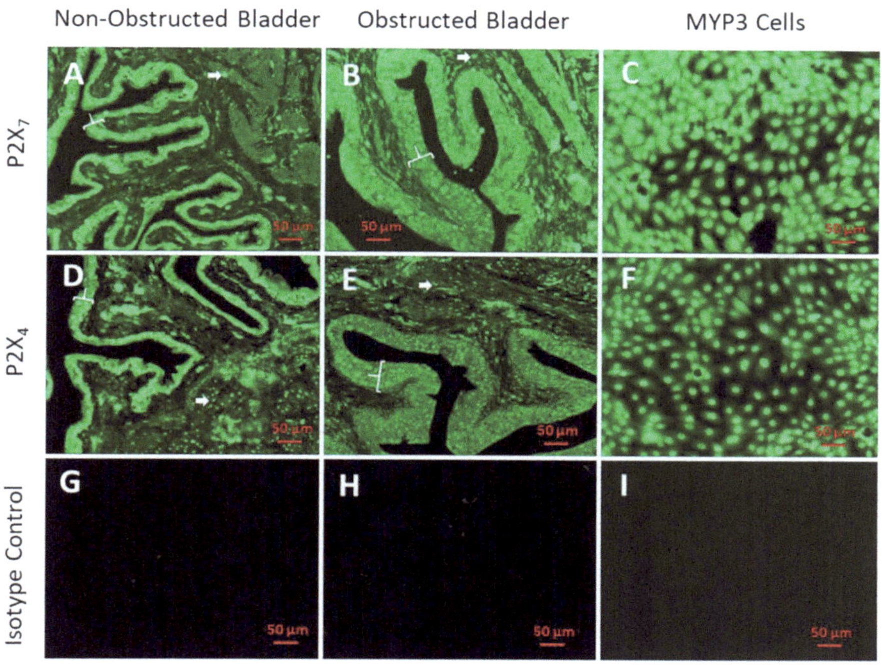

Fig. 3.21 Localization of P2X7 and P2X4 in rat bladder and MYP3 cells. Immunofluorescence staining was performed on bladder tissue sections from healthy and pBOO rats as well as cultured MYP3 cells using antibodies against P2X7 or P2X4. Normal rabbit serum was used as an isotype control. Brackets indicate the urothelial layer, and arrows identify smooth muscle regions. Images were acquired at 20× magnification. (Adapted with permission from *Int Urol Nephrol* (Springer, 2018))

(Fig. 3.23). Together, these results support a model in which hydrostatic pressure triggers ATP release via vesicular and channel-mediated pathways, amplifies signaling through P2X7 receptors, and ultimately activates NLRP3 inflammasome and caspase-1 via P2X4 receptors (Fig. 3.24).

3.6.3 Discussion

The overarching objective of this work was to explore mechanisms by which pBOO may lead to inflammation within the bladder wall (Metcalfe et al. 2010). We proposed that sustained elevations in intravesical pressure—whether from impaired compliance during urine storage or from voiding against an obstruction—provoke extracellular ATP release, which in turn activates purinergic signaling pathways leading to caspase-1 activation in urothelial cells. To evaluate this hypothesis, MYP3 urothelial cells were subjected to hydrostatic pressures mimicking abnormal

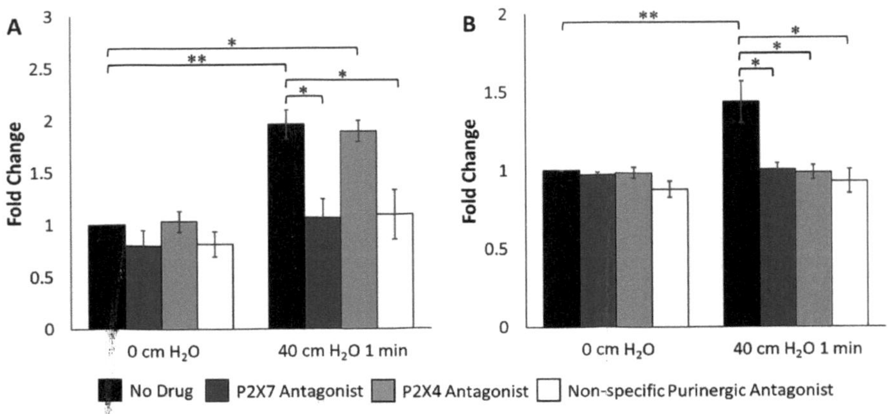

Fig. 3.22 Purinergic receptor antagonism reduces pressure-induced ATP release and caspase-1 activation. MYP3 cells were pretreated for 60 min with A-438079 (10 μM), 5-BDBD (20 μM), PPADS (25 μM), or no drug before exposure to hydrostatic pressure (40 cmH₂O, 1 min). Media were then collected for extracellular ATP quantification (**a**), and cell lysates were analyzed for caspase-1 activity (**b**). Data are expressed as fold change relative to the 0 cmH₂O no-drug control. Results are presented as mean ± SEM, with statistical analysis by one-way ANOVA and Tukey's post hoc test. *$P < 0.05$, **$P < 0.01$, $n = 8$. (Adapted with permission from *Int Urol Nephrol* (Springer, 2018))

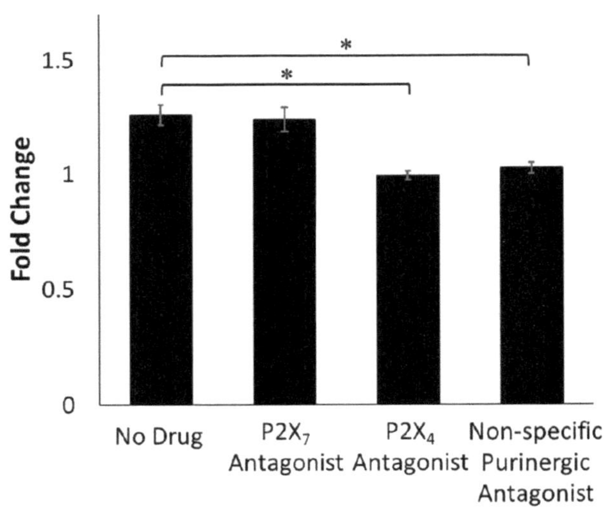

Fig. 3.23 ATP-stimulated caspase-1 activity in MYP3 cells. Cells were pretreated for 60 min with A-438079 (10 μM), 5-BDBD (20 μM), PPADS (25 μM), or no drug, and then incubated with 500 μM ATP disodium salt solution for 60 min. Caspase-1 activity was measured in cell lysates and expressed as fold change relative to the 0 μM ATP, no-drug control. Results are shown as mean ± SEM on a $\log_{10}$ scale. *$P < 0.05$, $n = 6$. (Adapted with permission from *Int Urol Nephrol* (Springer, 2018))

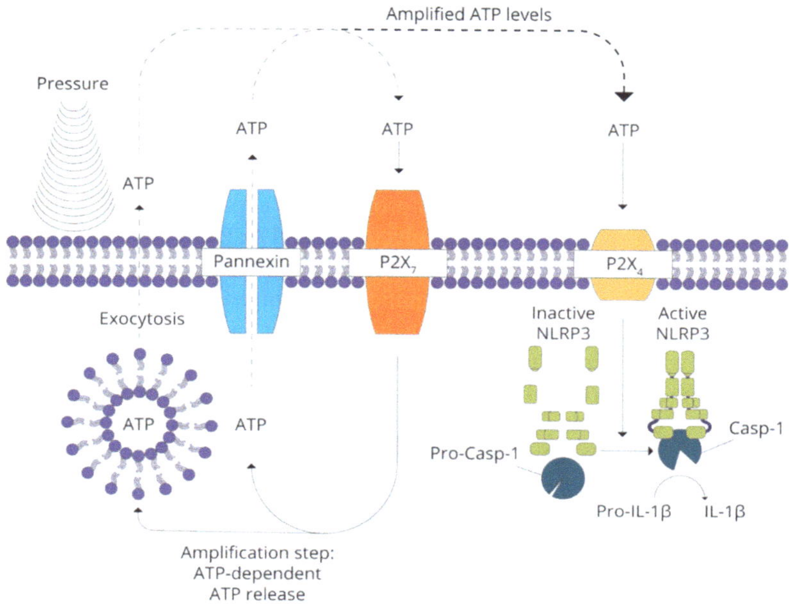

Fig. 3.24 Proposed mechanism of pressure-induced ATP release and NLRP3 inflammasome activation. Hydrostatic pressure initiates ATP release from urothelial cells through pannexin-1 hemichannels and vesicular exocytosis. Extracellular ATP then engages P2X7 receptors, amplifying ATP release and further elevating extracellular ATP levels. The amplified ATP signal subsequently activates P2X4 receptors, which mediate NLRP3 inflammasome activation and downstream caspase-1–dependent inflammatory responses. (Adapted with permission from *Int Urol Nephrol* (Springer, 2018))

storage (15 cmH_2O for 60 min) and obstructed voiding (40 cmH_2O for 1 min). Both pressure regimens, as well as hypotonic swelling, resulted in elevated extracellular ATP concentrations and increased intracellular caspase-1 activity (Fig. 3.19). While earlier studies had reported that cell swelling alone could induce ATP release from urothelial cells (Birder et al. 2002; Boudreault and Grygorczyk 2004), our data extend these findings by demonstrating that hypotonic stress is also sufficient to stimulate caspase-1 activation (Fig. 3.19b). Together with prior reports showing that physiologic levels of hydrostatic pressure can induce ATP release in urothelial cells (Olsen et al. 2011), retinal ganglion cells (Reigada et al. 2008), and lung epithelium (Richter et al. 2014), these results suggest that acute increases in pressure can rapidly initiate not only ATP release but also inflammasome-related responses in bladder urothelium.

ATP is recognized as a DAMP capable of triggering inflammasome activation, and our findings provide evidence that hydrostatic pressure–induced caspase-1 activation is mediated by multiple purinergic receptor subtypes functioning in sequence. Broad inhibition of purinergic signaling with PPADS reduced both ATP release and caspase-1 activity (Fig. 3.22), supporting a receptor-mediated mechanism. More specifically, antagonism of P2X7 with A-438079 attenuated pressure-induced ATP release as well as caspase-1 activation (Fig. 3.22), whereas inhibition of P2X4 with 5-BDBD selectively suppressed caspase-1 activity without affecting extracellular ATP levels (Fig. 3.22). These distinct effects suggest that P2X7 is critical for initiating ATP-dependent ATP release, while P2X4 participates in downstream NLRP3 inflammasome activation.

Further support for this model was provided by experiments using exogenous ATP stimulation. Exposure of MYP3 cells to 500 μM ATP elevated caspase-1 activity, and this response was not altered by P2X7 antagonism. In contrast, blockade of P2X4 or broad purinergic inhibition significantly diminished ATP-induced caspase-1 activity (Fig. 3.23). These observations indicate that once extracellular ATP levels are sufficiently high, signaling through P2X4 is the primary driver of inflammasome activation. This dual role of P2X7 and P2X4 helps reconcile earlier reports that NLRP3 activity can persist even in P2X7 knockout models (Franceschini et al. 2015).

Previous studies have emphasized P2X7 as the canonical activator of the NLRP3 inflammasome (Pelegrin and Surprenant 2006; Franceschini et al. 2015; Di Virgilio 2013). While our data are consistent with P2X7 playing an essential upstream role, they also implicate P2X4 as a necessary mediator of caspase-1 activation downstream of ATP amplification. We propose a mechanistic sequence in which hydrostatic pressure induces ATP release that acts initially on P2X7 to promote further ATP efflux, thereby amplifying the extracellular signal. This elevated ATP then engages P2X4 receptors to activate NLRP3 and caspase-1, resulting in inflammatory signaling (Fig. 3.24). The model is further supported by evidence from other systems, including ATP release in the retina under pressure (Reigada et al. 2008) and in bladder urothelium (Nakagomi et al. 2016). These findings carry broader implications since the NLRP3 inflammasome has been linked to inflammation across diverse tissues such as the lung (Grailer et al. 2014), cerebral cortex (Liu et al. 2013), and myocardium (Takahashi 2014). Thus, mechanically driven, purinergic-dependent activation of inflammasome pathways may represent a general paradigm extending beyond bladder pathophysiology.

In summary, this study demonstrates for the first time that elevated hydrostatic pressure can directly activate caspase-1 in urothelial cells through sequential involvement of P2X7 and P2X4 receptors. Based on this mechanism, chronic voiding against obstruction in pBOO could promote urothelial inflammation, fibrosis of the bladder wall, and progressive loss of compliance. Persistent exposure to elevated storage pressures would be expected to further aggravate these processes. Nonetheless, additional studies are needed to clarify the precise molecular events linking pressure-induced ATP release to long-term changes in bladder structure and function.

3.7 BOO Induces Fibrosis and EMT In Urothelial Cells Which Can Be Recapitulated In Vitro Through Elevated Storage and Voiding Pressure Cycles

3.7.1 Introduction

Previous investigations have demonstrated that bladder urothelial cells release ATP when subjected to elevated hydrostatic pressure. This ATP release has been proposed to play a role in normal physiology, such as detecting bladder fullness (Olsen et al. 2011), as well as in pathological states like bladder outlet obstruction (BOO) (Dunton et al. 2018). Most prior studies, however, have emphasized short-term responses occurring within minutes of pressure exposure, while the prolonged effects of such mechanical stress remain less understood.

BOO is known to progress through distinct stages characterized by changes in bladder pressure profiles. In the acute phase, an increase in voiding pressure (VP) predominates, reflecting the greater contractile force required to expel urine against the obstructed outlet (Bosch et al. 2019; Pétrilli et al. 2007; Lamkanfi et al. 2009; Hughes et al. 2016). In the chronic phase, both VP and storage pressure (SP) become elevated, likely as a result of reduced bladder wall compliance (Pétrilli et al. 2007; Lamkanfi et al. 2009; Iguchi et al. 2014; Islam et al. 2014). These pressure changes correspond with biochemical alterations such as urothelial and smooth muscle inflammation, which may eventually drive fibrosis. Indeed, animal studies have shown that pharmacological blockade of NLRP3 inflammasome activation reduces collagen accumulation in the bladder (Pétrilli et al. 2007; Lamkanfi et al. 2009; Hughes et al. 2016; Hughes et al. 2017), and in vitro work has demonstrated that brief exposure to high hydrostatic pressure (40 cmH_2O for 1 min) activates NLRP3 inflammasome signaling in urothelial cells (Dunton et al. 2018). Collectively, these findings suggest that the mechanical environments characteristic of acute versus chronic BOO may differentially regulate NLRP3-driven pathways contributing to fibrosis.

Although extracellular matrix (ECM) synthesis is not traditionally considered a function of epithelial cells, urothelial cells have been reported to express fibrotic markers such as collagen (Tekgul et al. 1996; Schröder et al. 2013), prolyl-4-hydroxylase (P4H), lysyl oxidase (LOX) (Iguchi et al. 2014; Hughes et al. 2017), and transforming growth factor beta (TGF-β) (Iguchi et al. 2014; Islam et al. 2014). This evidence points toward a potential epithelial-to-mesenchymal transition (EMT) contributing to fibrosis in BOO (Iguchi et al. 2014). EMT is defined by the loss of epithelial polarity and adhesion, coupled with increased ECM synthesis and remodeling (Iguchi et al. 2014; Islam et al. 2014; Sharif-Afshar et al. 2005; Baskin et al. 1996). On a molecular level, EMT is characterized by downregulation of epithelial proteins such as E-cadherin (ECAD) and cytokeratins (CK), with concurrent upregulation of mesenchymal markers including N-cadherin (NCAD) and α-smooth muscle actin (α-SMA) (Islam et al. 2014; Geng et al. 2015). In addition, elevated expression of matrix metallopeptidase-9 (MMP-9) and activation of TGF-β pathways have been implicated in driving EMT (Orlichenko and Radisky 2008). Supporting this, exposure of porcine urothelial cells to TGF-β reduced ECAD and CK levels while increasing NCAD, α-SMA, and MMP-9 expression (Islam et al. 2014). Despite these insights, the relationship between mechanically altered bladder pressures in BOO and urothelial expression of TGF-β or EMT-associated genes has not been clarified.

In the present study, the effects of acute and chronic BOO-like pressure cycles on molecular markers of EMT and fibrosis were investigated using both an in vivo rat model and an in vitro MYP3 urothelial cell model. Furthermore, because NLRP3 inflammasome activation has been implicated in BOO-induced fibrosis, the contribution of this pathway to collagen production in MYP3 cells was assessed in the presence and absence of NLRP3 inhibition.

3.7.2 Summary of Experimental Results

To investigate the role of purinergic receptors and inflammasome activation in bladder responses to hydrostatic pressure (HP), we employed both a rat model of partial bladder outlet obstruction (pBOO) (Hughes et al. 2016) and an established urothelial cell culture system (Cohen et al. 2002; Nascimento et al. 2008; Kawamata et al. 1993).

In vivo experiments: Four experimental groups were generated: (1) untreated control rats, (2) sham-operated rats in which a loose urethral ligature was placed, (3) BOO rats receiving vehicle, and (4) BOO rats receiving the NLRP3 inhibitor glyburide delivered via a 21-day slow-release pellet. Animals were sacrificed either at 14 days (acute phase) or 42 days (chronic phase), and urothelial cells were harvested from the excised bladders. Bladder explants from both acute and chronic BOO groups exhibited significant increases in bladder weight relative to controls (Fig. 3.25a). Glyburide treatment reduced bladder weight in the acute BOO animals, but no significant effect was seen in the chronic BOO group (Fig. 3.25a).

Gene expression analysis revealed that urothelial cells from acute BOO rats had elevated expression of fibrotic markers including Col I, Col III, and LOX (Fig. 3.25b), while in the chronic group, only Col I expression was significantly increased (Fig. 3.25c). Glyburide attenuated these changes in both acute and chronic groups. In parallel, epithelial markers ECAD and CK20 were downregulated, whereas mesenchymal markers αSMA, NCAD, and MMP-9 were upregulated in both acute and chronic BOO urothelial cells (Fig. 3.25d, e). Notably, glyburide treatment reversed

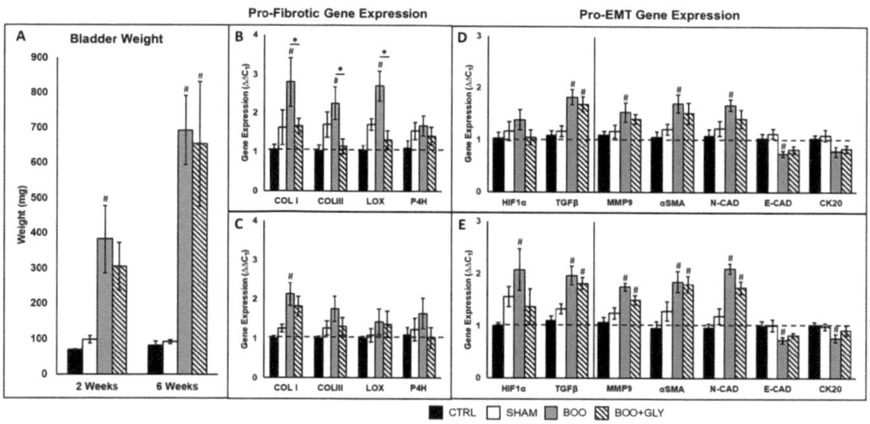

Fig. 3.25 Fibrotic and EMT markers in urothelial cells from acute and chronic BOO rats. Bladder weights increased in both acute (2 weeks) and chronic (6 weeks) BOO groups, with glyburide treatment reducing weight in acute but not chronic obstruction (**a**). Urothelial cells (UCs) from acute BOO rats displayed elevated expression of Col I, Col III, and LOX (**b**), whereas those from chronic BOO rats showed increased Col I expression only (**c**). Glyburide treatment reduced pro-fibrotic gene expression in both groups (B and C). UCs from both acute and chronic BOO rats exhibited decreased epithelial markers (E-CAD, CK20) and increased mesenchymal markers (MMP9, αSMA, N-CAD) (**d**, **e**). Glyburide had minimal effect on EMT gene expression (**d**, **e**). TGF-β was elevated in both phases of BOO regardless of treatment (**d**, **e**), while HIF1α expression was specifically elevated in chronic BOO and attenuated by glyburide (**e**). Data are mean ± SEM analyzed by one-way ANOVA with Tukey post-hoc test. # indicates comparison to control + no-drug treatment; * indicates between-group comparisons; $p < 0.05$, $n = 6$. (Adapted with permission from *Int Urol Nephrol*, Springer (2021))

these EMT-associated changes in the acute group but was ineffective in the chronic group.

In vitro experiments: MYP3 cells (a non-tumorigenic immortalized rat urothelial line (Cohen et al. 2002; Nascimento et al. 2008; Kawamata et al. 1993)) were exposed to pressure cycles for 72 h using a custom HP apparatus (Stover and Nagatomi 2007). Pressure regimens were designed to mimic: (1) unobstructed bladder (0 cmH2O for 175 min + 30 cmH_2O for 5 min), (2) acute BOO (0 cmH_2O for 175 min + 75 cmH_2O for 5 min), and (3) chronic BOO (15 cmH_2O for 175 min + 75 cmH_2O for 5 min). Acute and chronic regimens were introduced sequentially after an initial 24-h exposure to unobstructed conditions, reflecting disease progression (Fig. 3.26). In some experiments, cells were pretreated with the caspase-1 inhibitor Ac-YVAD-cmk (50 μM, 60 min).

Exposure to both acute and chronic BOO-like pressure cycles significantly increased intracellular caspase-1 activity, consistent with NLRP3 inflammasome activation (Fig. 3.27a). Pretreatment with Ac-YVAD-cmk suppressed caspase-1 activity to control levels (Fig. 3.27a). Chronic BOO conditions also induced a significant rise in soluble collagen release into the medium, which was abolished by caspase-1 inhibition (Fig. 3.27b).

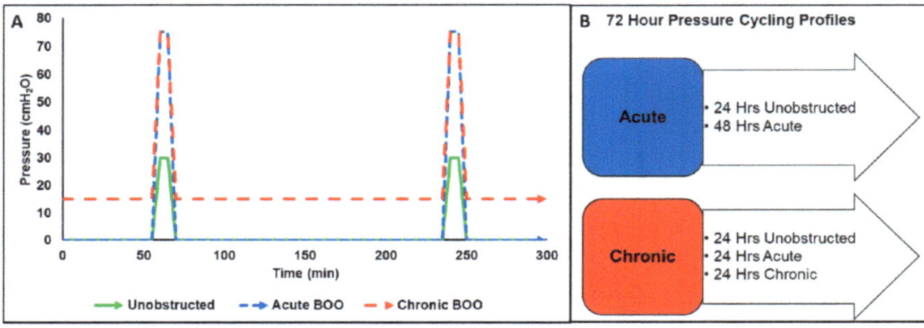

Fig. 3.26 Schematic of pressure profiles simulating acute and chronic BOO. Urothelial cells were exposed to defined pressure patterns replicating unobstructed (green), acute (blue), and chronic (red) BOO conditions (**a**). For 72-h experiments, acute BOO was modeled by 24 h of unobstructed conditions followed by 48 h of acute pressure (**b**). Chronic BOO was modeled by sequential exposure: 24 h unobstructed, 24 h acute, and 24 h chronic pressure (**b**). These regimens mimic the progression of BOO in vivo. (Adapted with permission from *Int Urol Nephrol*, Springer (2021))

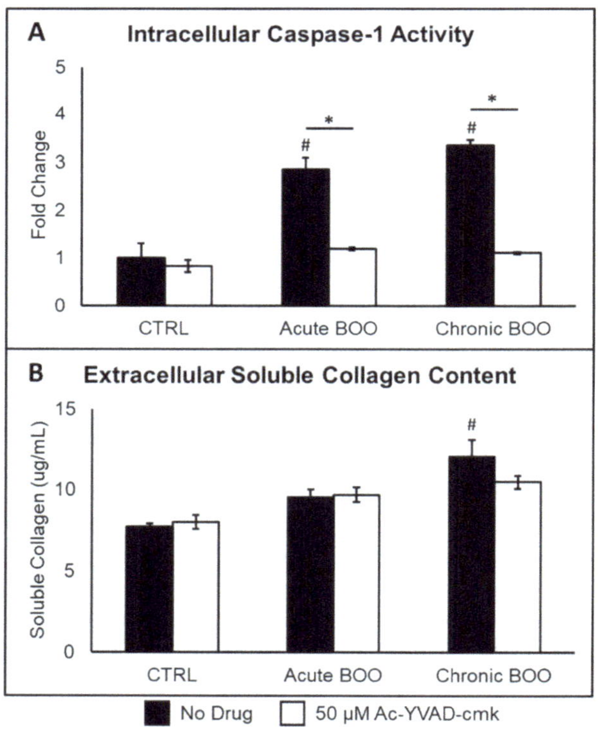

Fig. 3.27 Caspase-1 activation and collagen secretion under acute and chronic BOO pressure cycling. Caspase-1 activity increased after 72 h of acute or chronic pressure cycling, while treatment with Ac-YVAD-cmk reduced activity to control levels (**a**). Chronic pressure cycling increased soluble collagen release, which was suppressed by Ac-YVAD-cmk (**b**). Data are mean ± SEM analyzed using one-way ANOVA with Tukey post-hoc test. # indicates comparison to control + no-drug group; * indicates between-group differences; $p < 0.05$, $n = 4$. (Adapted with permission from *Int Urol Nephrol*, Springer (2021))

Regarding EMT-associated genes, acute and chronic BOO pressure regimes reduced expression of epithelial markers ECAD and CK20, although the decrease in ECAD was not statistically significant (Fig. 3.28a, b). Both regimens increased mesenchymal markers αSMA and NCAD as well as MMP-9 expression (Fig. 3.28a, b). In addition, TGF-β expression was elevated, while HIF1α levels were unchanged (Fig. 3.28c, d). Ac-YVAD-cmk treatment lowered αSMA, MMP-9, and TGF-β expression to control levels in the acute group, but had little effect under chronic conditions, where αSMA, NCAD, MMP-9, and TGF-β remained elevated (Fig. 3.28a, b, d).

Morphological analysis: Phase-contrast imaging revealed that MYP3 cells exposed to acute or chronic pressure regimens exhibited more elongated morphologies, reflected by increased aspect ratios compared to controls (Fig. 3.29a–c, g). A decreasing trend in average cell area was observed but did not reach statistical significance (Fig. 3.29h–n). Treatment with the caspase-1 inhibitor had minimal impact on either cell elongation or cell area (Fig. 3.29d–f, k–m).

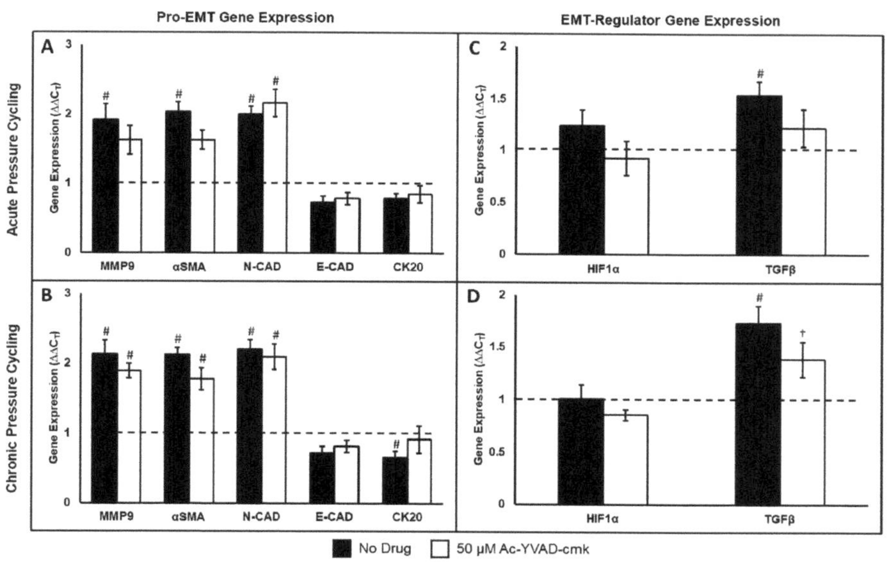

Fig. 3.28 EMT marker expression in urothelial cells exposed to pressure cycling. After 72 h, both acute and chronic pressure regimens elevated expression of mesenchymal markers MMP9, αSMA, and N-CAD, while CK20 expression decreased under chronic conditions (**a**, **b**). Ac-YVAD-cmk treatment reduced MMP9 and αSMA expression in the acute group (**a**) and restored CK20 in the chronic group (**b**). HIF1α expression was unchanged (**a**, **b**). TGF-β expression increased in both conditions, with Ac-YVAD-cmk attenuating this effect only under acute pressure (**c**, **d**). Data are mean ± SEM analyzed with one-way ANOVA and Tukey post-hoc test. # $p < 0.05$ vs. control (one-fold); † $p < 0.1$ vs. control (one-fold); $n = 4$. (Adapted with permission from *Int Urol Nephrol*, Springer (2021))

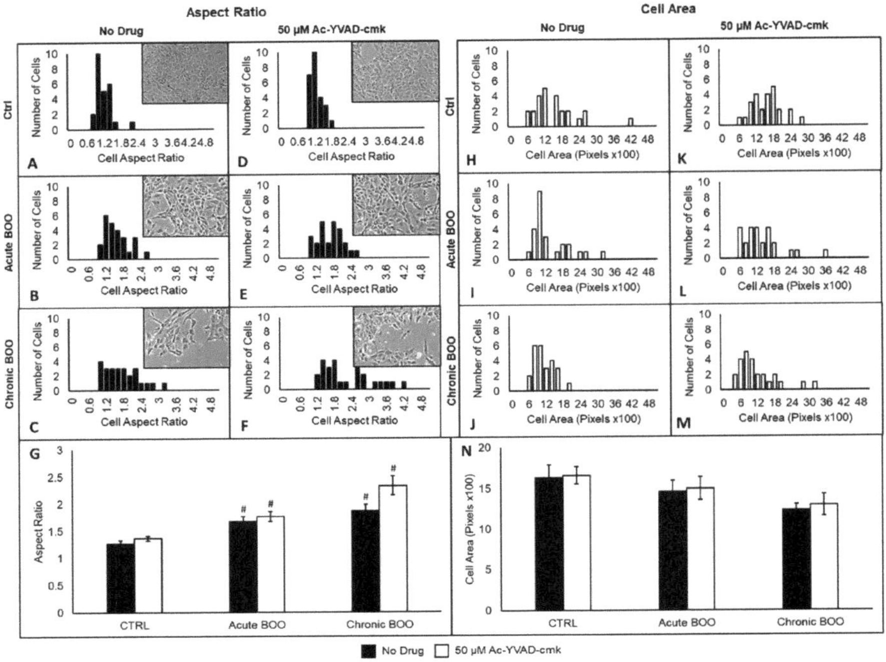

Fig. 3.29 Morphological changes in urothelial cells after pressure cycling. After 72 h, cells exposed to acute or chronic pressure cycling exhibited increased aspect ratios (**a–c**, **g**). Ac-YVAD-cmk did not alter this effect (**d–f**, **g**). Pressure cycling produced a downward trend in cell area, though not statistically significant (**h–j**, **n**). Treatment with Ac-YVAD-cmk did not affect aspect ratio or cell area (**k–m**, **n**). Bar graph data (**g**, **n**) represent mean ± SEM analyzed with one-way ANOVA and Tukey post-hoc test. # $p < 0.05$ vs. control + no-drug group; n = 25. (Adapted with permission from *Int Urol Nephrol*, Springer (2021))

3.7.3 Discussion

The present study aimed to clarify the cellular mechanisms driving bladder fibrosis in response to altered pressure cycling during BOO. We hypothesized that urothelial cells undergo epithelial-to-mesenchymal transition (EMT) in response to the distinct changes in voiding and storage pressure that accompany the progression from acute to chronic BOO. To test this, we examined EMT and fibrotic markers in urothelial cells isolated from rats at different stages of BOO and evaluated whether pressure cycling alone was sufficient to reproduce these changes in vitro.

Consistent with earlier work, the acute phase of BOO (<2 weeks) is characterized by elevations in voiding pressure (VP) of 30–75 cmH_2O, while storage pressure (SP) remains largely unchanged (Hughes et al. 2016). In the chronic phase (>5 weeks), both VP (30–75 cmH_2O) and SP (5–15 cmH_2O) are elevated relative to unobstructed controls (Hughes et al. 2019; Choo et al. 2017; Malkowicz et al. 1986). These physiological observations guided the design of our in vitro pressure regimens: unobstructed (0 cmH_2O SP; 30 cmH_2O VP), acute BOO (0 cmH_2O SP; 75 cmH_2O VP), and chronic BOO (15 cmH_2O SP; 75 cmH_2O VP).

Our findings demonstrate that the molecular responses of urothelial cells to BOO are phase-dependent. In rats, acute BOO was associated with increased expression of Col I, Col III, and LOX, whereas chronic BOO produced elevation in Col I alone. Similarly, MYP3 cells subjected to acute pressure cycles upregulated pro-fibrotic genes without detectable increases in collagen protein, while chronic pressure conditions resulted in both gene upregulation and elevated collagen protein levels. These data suggest that heightened voiding pressure in acute BOO initiates transcriptional activation of fibrotic pathways (Malkowicz et al. 1986; Metcalfe et al. 2010), whereas the superimposed increase in storage pressure during chronic BOO is necessary for translation and accumulation of extracellular matrix proteins such as collagen. Together with prior reports, our results highlight elevated storage pressure as a critical mechanical signal that drives fibrosis in the obstructed bladder.

Another important observation is that urothelial cells from BOO rats, as well as MYP3 cells exposed to pressure cycles, displayed hallmarks of EMT, including reduced expression of epithelial markers and increased expression of mesenchymal markers. Morphological changes in vitro further supported this transition. These results are consistent with prior reports in rats (Hughes et al. 2017; Kim et al. 2000; Kadam et al. 2021), mice (Andersson and Gratzke 2007), and rabbits (Schröder et al. 2013), which showed that BOO conditions promote EMT in the urothelium. However, our study provides new evidence that this process can be induced specifically through pressure-dependent cues.

The role of the NLRP3 inflammasome was also examined. Pharmacological inhibition of NLRP3 reduced the fibrotic gene response, supporting its involvement in BOO-associated fibrosis. However, suppression of NLRP3 did not fully block EMT-associated changes, particularly the upregulation of TGF-β. Given the established role of TGF-β in EMT across tissues (Islam et al. 2014), these results indicate that BOO-induced EMT involves both NLRP3-dependent and NLRP3-independent pathways. Thus, interventions targeting TGF-β signaling, such as blockade of Smad2/3, may represent promising strategies for mitigating EMT and fibrosis in BOO (Islam et al. 2014).

Finally, this study demonstrates that the hydrostatic pressure regimens developed to mimic phases of BOO can reproduce many of the cellular and molecular events observed in vivo, including EMT and fibrosis (Iguchi et al. 2014; Hughes et al. 2017; Tekgul et al. 1996). Nonetheless, the in vitro model lacks additional pathological stimuli present in obstructed bladders, such as chronic stretch, ischemia-induced hypoxia, and paracrine signals from neighboring cell types. Future studies incorporating co-culture systems and additional mechanical or chemical stressors will be important to more fully characterize the complex environment that drives BOO pathology.

3.8 Stretch and Pressure Differentially Trigger Bladder Urothelial Cell Responses

3.8.1 Introduction

Bladder outlet obstruction (BOO) is a prevalent urological condition, estimated to affect over one billion individuals globally (Irwin et al. 2011). It is characterized by inflammatory changes (Hughes et al. 2016; Hughes et al. 2019) and fibrotic remodeling (Hughes et al. 2017) of the bladder wall. To better understand the cellular and molecular basis of the inflammation and fibrosis observed in BOO patients, numerous in vitro studies have investigated how bladder cells respond to increased pressure and/or stretch (Yu et al. 2009; Du et al. 2007; Olsen et al. 2011; Kumar et al. 2004; Dunton et al. 2018; Yu 2015; Sano et al. 2016). Our prior work demonstrated that both primary rat urothelial cells (Sect. 3.5) (Olsen et al. 2011 and the immortalized MYP3 line (Sect. 3.6) (Dunton et al. 2018) release ATP in response to hydrostatic pressure, which in turn activates the NLRP3 inflammasome and drives inflammatory responses (Olsen et al. 2011; Dunton et al. 2018). Conversely, other reports have suggested that tissue stretch alone is the predominant mechanical cue responsible for calcium signaling, ATP release, and subsequent inflammation (Mochizuki et al. 2009; Sano et al. 2016; Lee et al. 2007). These divergent findings underscore the need to clarify the respective roles of stretch and pressure as mechanical stimuli that activate inflammasome pathways and contribute to bladder pathology. Mechanistic studies indicate that urothelial ATP release during stretch may occur through several distinct pathways including vesicular exocytosis (VE) (Yu et al. 2009; Lee et al. 2007) and the involvement of hemi-channel proteins, including pannexin-1 (Panx1) (Beckel et al. 2015) and connexin-43 (Cx43) (Sengiku et al. 2018). In many of these studies, stretch was induced by applying differential pressure, making it difficult to fully separate the contributions of stretch versus hydrostatic load. It is therefore plausible that the same ATP release mechanisms also function during pressure-induced signaling, although this has yet to be demonstrated.

A key component of mechanotransduction is the rise in intracellular calcium, which links physical cues to downstream biochemical events. Stretch of isolated murine urothelial cells (20–50%) has been shown to increase intracellular Ca^{2+} concentration (Mochizuki et al. 2009; Miyamoto et al. 2014; Lee et al. 2007). Moreover, several calcium-permeable ion channels expressed in the urothelium (Mochizuki et al. 2009; Miyamoto et al. 2014; Yamada et al. 2009; Shabir et al. 2013; Michishita et al. 1973) have been implicated in mechanosensation, including members of the transient receptor potential (TRP) family (Mochizuki et al. 2009; Suzuki et al. 2003; Servin-Vences et al. 2017) and, more recently, the mechanosensitive channel Piezo1 (Miyamoto et al. 2014; Servin-Vences et al. 2017; Li et al. 2015). These reports support roles for TRPV4 and Piezo1 in detecting stretch, but no definitive pressure-sensing mechanism has been identified in bladder urothelium. Interestingly, Piezo1 has been linked to pressure-induced ATP release in other tissues such as chondrocytes (Servin-Vences et al. 2017) and cancer cells (Li et al. 2015), raising the possibility of a similar role in bladder cells. Based on this evidence, we propose that both stretch and pressure serve as distinct mechanical cues, engaging different sensory pathways but converging on ATP release and inflammation in urothelial cells.

In the present work, we directly compared how MYP3 cells respond to stretch versus hydrostatic pressure using a custom-designed bioreactor system. Cells were subjected to mechanical regimes mimicking bladder filling, storage, and voiding cycles. We then quantified ATP release and inflammasome activation in the presence and absence of pharmacological inhibitors targeting specific channels, receptors, and signaling molecules. This approach was designed to delineate the mechanosensory ion channels and ATP release pathways by which stretch and/or pressure activate the NLRP3 inflammasome under BOO-like conditions.

3.8.2 Summary of Experimental Results

To delineate how urothelial cells respond to different mechanical cues, we designed a custom bi-compartment bioreactor capable of independently applying either hydrostatic pressure or membrane stretch, or both simultaneously. As illustrated in Fig. 3.30, the device consisted of two chambers separated by a silicone membrane seeded with MYP3 cells. Each chamber was fitted with ports that allowed for controlled fluid/gas exchange and precise pressure regulation. Stretch was induced by generating negative pressure in the lower chamber, thereby deforming the silicone membrane upward. Hydrostatic pressure, in contrast, was applied by pressurizing the gas phase above the culture media in the upper chamber. The lower chamber was filled with water to eliminate unintended stretching of the membrane during pressurization, ensuring that stretch and pressure could be tested independently.

The silicone substrates were mechanical characterized to quantify strain transmission to the adherent cells. Biaxial tensile testing was performed for eight loading cycles (up to 35% strain at 2% strain/s). Stress ($\sigma = F/A$) and stretch ($\lambda = (\Delta L + L_0)/L_0$) values from the final three cycles were analyzed. The stress–strain data from the eighth cycle were fitted to a five-parameter Mooney–Rivlin hyperelastic model, which was subsequently used in finite-element simulations of membrane deformation. Radial strains were calculated using arc length integration:

$$L = \int_a^b \sqrt{1 + \left(\frac{dy}{dx}\right)^2}\, dx;$$

where *a* and *b* represent radial positions from the edge to the center of the substrate, and *dy/dx* describes curvature in the z-direction. From this, average radial and circumferential strains were determined.

Biaxial testing showed that after six cycles the stress–stretch relationship stabilized, indicating mechanical preconditioning (Fig. 3.31a, b). Stress–stretch behavior was consistent in axial and transverse directions, with evident hysteresis between loading and unloading curves. Computational modeling yielded a close fit

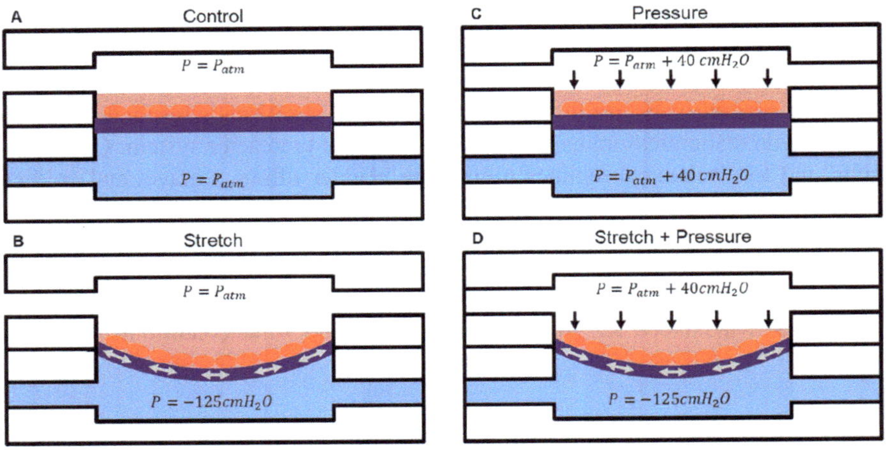

Fig. 3.30 Schematic of the custom pressure–stretch bioreactor. The lower chamber was filled with PBS (blue) to enable negative pressure application, while culture medium (red) was added to the upper chamber to sustain the urothelial cells seeded on the silicone substrate. **Control (CTRL):** upper chamber open to atmosphere, no negative pressure applied to the lower chamber (**a**). **Stretch (S):** substrate deformation achieved by applying −125 cmH_2O negative pressure to the lower chamber with the upper chamber vented to atmosphere (**b**). **Pressure (P):** 40 cmH_2O hydrostatic pressure applied to the upper chamber while the lower chamber remained sealed and incompressible, preventing substrate deformation (**c**). **Stretch + Pressure (S + P):** negative pressure (−125 cmH_2O) first applied to the lower chamber to stretch the substrate, followed by 40 cmH_2O applied to the upper chamber (**d**). (Adapted from Clemson University PhD Dissertation (2020))

(RMSE = 7.11; Fig. 3.31c). Model predictions indicated that pressures of 125, 175, and 225 cmH_2O corresponded to average substrate strains of 0.16, 0.19, and 0.26, with maximum strain values of 0.20, 0.31, and 0.40, respectively, and minimum strain values of 0.07, 0.08, and 0.18 (Fig. 3.31d).

MYP3 cells (immortalized rat urothelial line (Kawamata et al. 1993; Nascimento et al. 2008)) were seeded on gelatin-coated silicone membranes and mounted in the bioreactor with the cell surface facing upward. Cells were exposed for 15 min to one of four experimental conditions:

1. **CTRL**: no stretch, atmospheric pressure only.
2. **Stretch (S)**: strain ramped at 3%/min for 5 min, then held at 15% strain for 5 min.
3. **Pressure (P)**: exposed to 40 cmH_2O hydrostatic pressure for 1 min following 14 min at baseline.
4. **Stretch + Pressure (S + P)**: combination of conditions 2 and 3.

Supernatants were collected for ATP assays, and cell lysates analyzed for caspase-1 activity (Dunton et al. 2018). All conditions were repeated in the presence or absence of pharmacological inhibitors (Table 3.2), added 60 min prior to mechanical stimulation and maintained throughout experiments.

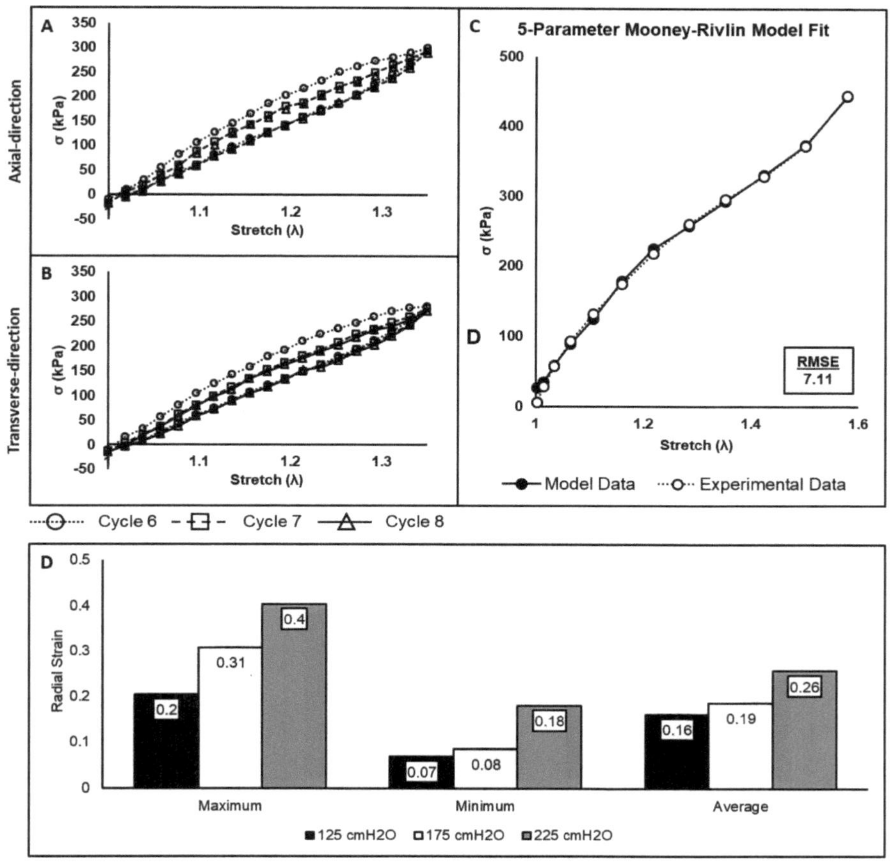

Fig. 3.31 Experimental and computational analysis of silicone substrate mechanics. Cyclic planar biaxial testing of the silicone substrate under prescribed equibiaxial strain demonstrated reproducible stress–stretch behavior by the seventh and eighth cycles in both the axial and transverse directions (**a**, **b**). A five-parameter Mooney–Rivlin model provided an accurate fit of the experimental stress–strain data obtained from planar biaxial testing, with a low root mean square error (RMSE) (**c**). Model predictions indicated that application of −125 cmH_2O generated an average radial strain of ~15%, with peak strain below 20% (**d**). (Adapted from Clemson University PhD Dissertation (2020))

Both stretch and pressure significantly increased ATP release relative to control, with the combined S + P condition producing the highest extracellular ATP levels (Fig. 3.32a). Caspase-1 activity was also elevated under S, P, and S + P compared to CTRL, though no synergistic increase was observed in the combined condition (Fig. 3.32b). Brefeldin A reduced ATP release and caspase-1 activity across all conditions (Fig. 3.33a, b). Panx1 inhibition with BB-FCF selectively reduced ATP

Table 3.2 Pharmacological inhibitors used for mechanosensing, ATP release, Ca^{2+} signaling, and vesicular exocytosis

Drug	Concentration	Target/Action
Brilliant blue FCF (BB-FCF)	10 μM	Pannexin-1 antagonist (Beckel et al. 2015)
Brefeldin A	10 μg/mL	Vesicular exocytosis inhibitor (Lee et al. 2007)
Probenecid	100 μM	Non-specific purinergic & Pannexin antagonist (Chopra et al. 2008)
GSK 2193874	1 μM	TRPV4 antagonist (Cheung et al. 2017)
BAPTA-AM	5 mM	Intracellular Ca^{2+} chelator (Birder et al. 2002)
TAT-GAP19	200 μM	Connexin-43 blocker (Li et al. 2018)
GsMTx4	20 μM	Piezo1 inhibitor (Miyamoto et al. 2014)

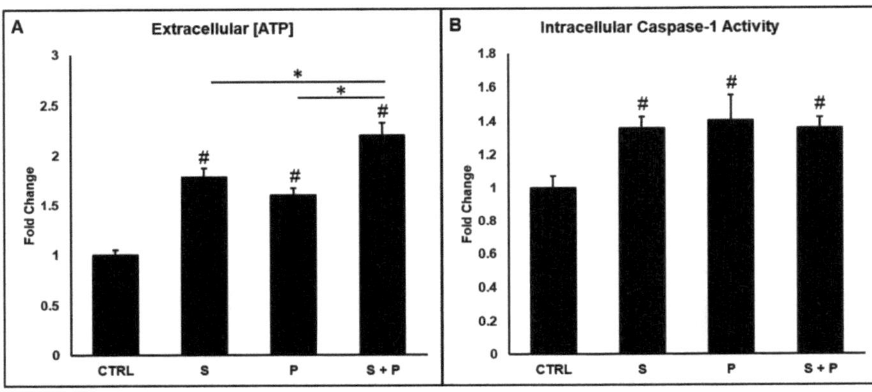

Fig. 3.32 Stretch and pressure independently stimulate ATP release, with additive effects, but not additive caspase-1 activation. MYP3 cells exhibited elevated extracellular ATP following exposure to stretch (S), pressure (P), or both (S + P). The combined condition (S + P) produced greater ATP release than either stimulus alone (**a**). Caspase-1 activity also increased with S, P, or S + P compared to control, though combined stimulation did not produce further enhancement beyond single stimuli (**b**). Data are presented as mean ± SEM and analyzed using one-way ANOVA with Tukey's post hoc test. Symbols: # $p < 0.05$ versus control; * $p < 0.05$ between groups. Sample sizes: n = 18 (CTRL). (Adapted from Clemson University PhD Dissertation (2020))

release and caspase-1 activation in pressure-only (P) conditions, but not under stretch. In contrast, blocking Cx43 with TAT-GAP19 reduced ATP release and caspase-1 activity under stretch (S, S + P) but not under pressure (P). Inhibition of TRPV4 or Piezo1 diminished ATP release and caspase-1 activity under stretch and S + P conditions but had no effect on pressure-only responses (Fig. 3.34a, b). Chelation of intracellular Ca^{2+} with BAPTA-AM reduced ATP release and caspase-1 activation across all conditions, including CTRL, suggesting that intracellular calcium flux is a necessary mediator of both pressure-and stretch-induced responses.

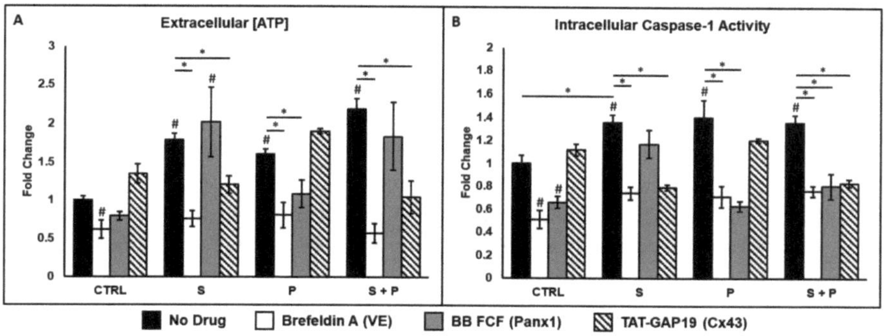

Fig. 3.33 Distinct ATP release mechanisms contribute to stretch-and pressure-induced responses. Pretreatment with brefeldin A (exocytosis inhibitor) or TAT-GAP19 (connexin-43 blocker) reduced ATP release in response to stretch (S) and stretch + pressure (S + P). Inhibition of pannexin-1 with Brilliant Blue FCF (BB-FCF) selectively reduced ATP release under pressure (P) conditions (**a**). Caspase-1 activity mirrored ATP release profiles: brefeldin A suppressed activity across all stimuli; BB-FCF reduced responses to pressure and combined stimuli; and TAT-GAP19 attenuated activity under stretch and S + P (**b**). Data are mean ± SEM; analysis performed with one-way ANOVA and Tukey's post hoc test. Symbols: # $p < 0.05$ versus control + no drug; * $p < 0.05$ between groups. Sample sizes: n = 18 (CTRL), n = 9 (drug-treated). (Adapted from Clemson University PhD Dissertation (2020))

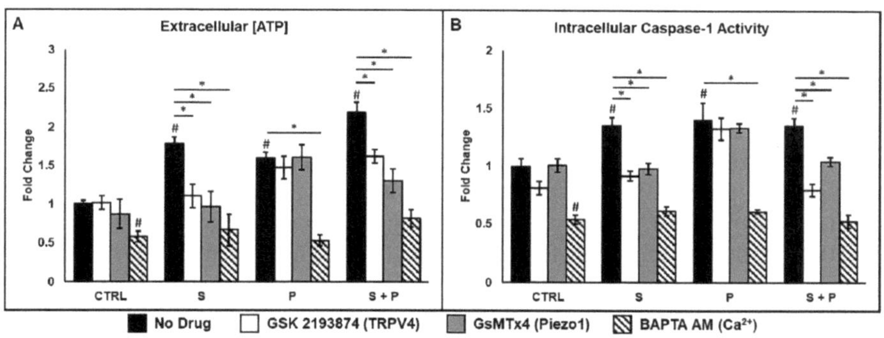

Fig. 3.34 TRPV4 and Piezo1 mediate stretch sensing, while intracellular Ca^{2+} is required for all mechanical responses. Pretreatment of MYP3 cells with TRPV4 antagonist (GSK 2193874) or Piezo1 inhibitor (GsMTx4) significantly reduced ATP release under stretch (S) and combined (S + P) conditions, but not during pressure (P) exposure (**a**). Caspase-1 activity was similarly attenuated under stretch-related conditions by TRPV4 and Piezo1 inhibition (**b**). In contrast, treatment with the intracellular Ca^{2+} chelator BAPTA-AM reduced both ATP release and caspase-1 activity across all conditions (CTRL, S, P, S + P) (B). Data are mean ± SEM; analysis performed with one-way ANOVA and Tukey's post hoc test. Symbols: # $p < 0.05$ versus control + no drug; * $p < 0.05$ between groups. Sample sizes: n = 18 (CTRL), n = 6 (drug-treated). (Adapted from Clemson University PhD Dissertation (2020))

3.8.3 Discussion

Because the bladder wall is subjected to both pressure and stretch during filling and voiding, it is important to clarify how these mechanical forces influence urothelial signaling pathways that drive ATP release and inflammasome activation. In this study, we hypothesized that under conditions mimicking BOO, pressure and stretch would each trigger ATP release but through distinct mechanisms. To directly test this, we constructed and calibrated a custom bioreactor capable of independently or jointly applying these mechanical cues.

To ensure accurate quantification of the strains imparted to the cells, biaxial tensile testing and computational modeling of the silicone substrate were performed. The material displayed viscoelastic deformation with notable hysteresis, consistent with prior descriptions of bladder tissue biomechanics (Nagatomi et al. 2004). Stress–stretch profiles were isotropic across orthogonal directions. The data were well represented by a five-parameter Mooney–Rivlin model, confirming that the computational predictions closely matched experimental behavior. When the lower chamber was subjected to −125 cmH_2O, the average radial strain was calculated at ~16%, a value comparable to distension observed in murine models of BOO (Chen et al. 2013b).

Cells were exposed to conditions designed to approximate in vivo loading. Stretch was applied prior to pressure in S + P conditions, reflecting bladder filling before voiding. The magnitude and rate of deformation were carefully chosen. Prior animal and ex vivo studies report bladder wall strains between 5–35% across physiological pressure ranges (10–80 cmH_2O) (Nagatomi et al. 2008; Morales-Orcajo et al. 2018), and rodent urothelial cells release ATP in response to ~20% stretch with concomitant rises in intracellular Ca^{2+} (Sun et al. 2001; Miyamoto et al. 2014; Lee et al. 2007). However, many of these studies used strain rates that far exceed physiological filling (10–40% per second (Sun et al. 2001; Miyamoto et al. 2014; Lee et al. 2007)), while natural bladder filling corresponds to ~4–7% per hour (≈0.0011–0.0019% per second) (Nagle et al. 2017). The present bioreactor applied stretch at ~3% per minute (0.05% per second), which is still elevated but represents a closer approximation than traditional models. For pressure, ~40 cmH_2O above baseline was selected, in line with pressure-volume relations measured during voiding in BOO rats (~70 cmH_2O peak pressure) (Lee et al. 2007).

Application of stretch and pressure together (S + P) elicited a stronger ATP release response than either stimulus alone (Fig. 3.33). This represents the first demonstration that physiological-rate loading can drive ATP release from urothelial cells in vitro. The additive nature of these responses suggests that stretch and pressure activate separate but complementary pathways. Pharmacological studies helped distinguish these mechanisms. Vesicular exocytosis was necessary for both stretch- and pressure-induced ATP release, as inhibition with brefeldin A reduced responses in all conditions. However, pathway-specific contributions were also evident: blocking pannexin-1 reduced ATP release and caspase-1 activity in pressure-stimulated cells, while connexin-43 inhibition attenuated responses under stretch but not pressure. Interestingly, when both forces were applied simultaneously, inhibition of Panx1 did not significantly reduce ATP release, implying that stretch-induced signaling via Cx43 and exocytosis dominates under combined stimulation.

Calcium influx has long been implicated in mechanically induced ATP release (Mochizuki et al. 2009; Miyamoto et al. 2014; McLatchie and Fry 2015). Consistent with this, chelation of intracellular Ca^{2+} suppressed both ATP release and inflammasome activation under all experimental conditions. Inhibition of TRPV4 or Piezo1 reduced stretch-induced responses but had no effect on pressure-induced ones, supporting their role as stretch sensors in the urothelium. These results suggest that activation of TRPV4 and Piezo1 elevates intracellular Ca^{2+}, which in turn promotes ATP release via exocytosis or Cx43 hemichannels. Moreover, Cx43 gap junctions could propagate Ca^{2+} signals to neighboring cells, amplifying inflammation across the epithelium (Toyofuku et al. 1998).

While stretch clearly engages TRPV4 and Piezo1, their role in pressure sensing remains unresolved. In macrophages and endothelial cells, Piezo1 has been shown to mediate pressure-induced pro-inflammatory and barrier-disrupting responses (Friedrich et al. 2019; Solis et al. 2019). However, in the present study, static pressure did not require TRPV4 or Piezo1 for ATP release or inflammasome activation. This aligns with prior work showing that macrophages exposed to static pressure alone exhibited little response (Solis et al. 2019), suggesting that Piezo1 may be preferentially activated by cyclic rather than constant pressure. Thus, although our results do not implicate Piezo1 in static pressure sensing in urothelial cells, it remains possible that cyclic hydrostatic loading could reveal a role for this channel.

Overall, this study demonstrates that stretch and pressure are both physiologically relevant stimuli in BOO, but they act through distinct signaling pathways to trigger ATP release and inflammasome activation. Importantly, our findings highlight that vesicular exocytosis is a shared pathway, while Panx1 and Cx43 represent pressure-and stretch-specific mechanisms, respectively. TRPV4 and Piezo1 serve as stretch sensors linked to Ca^{2+}-dependent ATP release. These insights underscore the importance of distinguishing between mechanical cues when developing pharmacological strategies aimed at mitigating inflammation and fibrosis in BOO.

References

Abdel-Karim AM, Abdel-Gawad M, Huynh H, Elhilali MM (2002) Modulation of insulin-like growth factor-I system of the bladder using a somatostatin analogue in chronic spinalized rats. J Urol 168(3):1253–1258. https://doi.org/10.1016/S0022-5347(05)64635-6

Acevedo AD, Bowser SS, Gerritsen ME, Bizios R (1993) Morphological and proliferative responses of endothelial cells to hydrostatic pressure: role of fibroblast growth factor. J Cell Physiol 157(3):603–614. https://api.istex.fr/ark:/67375/WNG-477MB61D-8/fulltext.pdf. https://doi.org/10.1002/jcp.1041570321

Adam RM, Kim J, Lin J et al (2002) Heparin-binding epidermal growth factor-like growth factor stimulates androgen-independent prostate tumor growth and antagonizes androgen receptor function. Endocrinology 143(12):4599–4608. https://doi.org/10.1210/en.2002-220561

Adam RM, Roth JA, Cheng H et al (2003) Signaling through PI3K/akt mediates stretch and PDGF-BB-dependent DNA synthesis in bladder smooth muscle cells. J Urol 169(6):2388–2393. https://doi.org/10.1097/01.ju.0000063980.99368.35
Adam RM, Eaton SH, Estrada C et al (2004) Mechanical stretch is a highly selective regulator of gene expression in human bladder smooth muscle cells. Physiol Genomics 20(1):36–44. https://doi.org/10.1152/physiolgenomics.00181.2004
Adelöw CAM, Frey P (2007) Synthetic hydrogel matrices for guided bladder tissue regeneration. In: Methods in molecular medicine, vol 140. Humana Press, Totowa, NJ, pp 125–140. https://doi.org/10.1007/978-1-59745-443-8_7
Alali AS, Fowler RA, Mainprize TG et al (2013) Intracranial pressure monitoring in severe traumatic brain injury: results from the american college of surgeons trauma quality improvement program. J Neurotrauma 30(20):1737–1746
Almasieh M, Wilson AM, Morquette B, Cueva Vargas JL, Di Polo A (2012) The molecular basis of retinal ganglion cell death in glaucoma. Prog Retin Eye Res 31(2):152–181
Andersson K, Gratzke C (2007) Pharmacology of α1-adrenoceptor antagonists in the lower urinary tract and central nervous system. Nat Clin Pract Urol 4(7):368–378. https://doi.org/10.1038/ncpuro0836
Apodaca G (2004) The uroepithelium: not just a passive barrier. Traffic (Copenhagen, Denmark) 5(3):117–128. https://doi.org/10.1046/j.1600-0854.2003.00156.x
Backhaus BO, Kaefer M, Haberstroh KM et al (2002) Alterations in the molecular determinants of bladder compliance at hydrostatic pressures less than 40 cm. H_2O. J Urol 168(6):2600–2604. https://doi.org/10.1016/S0022-5347(05)64226-7
Bar-Kochba E, Scimone MT, Estrada JB, Franck C (2016) Strain and rate-dependent neuronal injury in a 3D in vitro compression model of traumatic brain injury. Sci Rep 6(1):30550
Baskin LS, Hayward SW, Young P, Cunha GR (1996) Role of mesenchymal-epithelial interactions in normal bladder development. J Urol 156(5):1820–1827. https://doi.org/10.1016/S0022-5347(01)65545-9
Beckel JM, Daugherty SL, Tyagi P et al (2015) Pannexin 1 channels mediate the release of ATP into the lumen of the rat urinary bladder. J Physiol 593(8):1857–1871. https://doi.org/10.1113/jphysiol.2014.283119
Birder LA (2005) More than just a barrier: Urothelium as a drug target for urinary bladder pain. Am J Physiol Renal Physiol 289(3):489. https://doi.org/10.1152/ajprenal.00467.2004
Birder LA, Nakamura Y, Kiss S et al (2002) Altered urinary bladder function in mice lacking the vanilloid receptor TRPV1. Nat Neurosci 5(9):856–860. https://doi.org/10.1038/nn902
Birder LA, Barrick SR, Roppolo JR et al (2003) Feline interstitial cystitis results in mechanical hypersensitivity and altered ATP release from bladder urothelium. Am J Physiol—Ren Physiol 285(3):423–429. http://ajprenal.physiology.org/content/285/3/F423. https://doi.org/10.1152/ajprenal.00056.2003
Bommakanti RK, Vinayak S, Simonds WF (2000) Dual regulation of akt/protein kinase B by heterotrimeric G protein subunits. J Biol Chem 275(49):38870–38876. http://www.jbc.org/content/275/49/38870.abstract. https://doi.org/10.1074/jbc.M007403200
Bonacci JV, Stewart AG (2006) Regulation of human airway mesenchymal cell proliferation by glucocorticoids and beta2-adrenoceptor agonists. Pulm Pharmacol Ther 19(1):32–38. https://doi.org/10.1016/j.pupt.2005.02.011
Bonacci JV, Harris T, Stewart AG (2003) Impact of extracellular matrix and strain on proliferation of bovine airway smooth muscle. Clin Exp Pharmacol Physiol 30(5–6):324–328. https://doi.org/10.1046/j.1440-1681.2003.03838.x
Bosch R, Abrams P, Averbeck MA et al (2019) Do functional changes occur in the bladder due to bladder outlet obstruction? - ICI-RS 2018. Neurourol Urodyn 38(S5):S56–S65. https://doi.org/10.1002/nau.24076
Boudreault F, Grygorczyk R (2004) Cell swelling-induced ATP release is tightly dependent on intracellular calcium elevations. J Physiol 561(2):499–513. http://jp.physoc.org/content/561/2/499.abstract. https://doi.org/10.1113/jphysiol.2004.072306

Brazil DP, Hemmings BA (2001) Ten years of protein kinase B signalling: a hard akt to follow. Trends Biochem Sci 26(11):657–664

Caeyenberghs K, Leemans A, Leunissen I et al (2014) Altered structural networks and executive deficits in traumatic brain injury patients. Brain Struct Funct 219(1):193–209

Casella EM, Thomas TC, Vanino DL et al (2014) Traumatic brain injury alters long-term hippocampal neuron morphology in juvenile, but not immature, rats. Childs Nerv Syst 30(8):1333–1342

Cauwels A, Rogge E, Vandendriessche B, Shiva S, Brouckaert P (2014) Extracellular ATP drives systemic inflammation, tissue damage and mortality. Cell Death Dis 5(3):e1102. https://www.ncbi.nlm.nih.gov/pubmed/24603330. https://doi.org/10.1038/cddis.2014.70

Cavanaugh KJ, Cohen TS, Margulies SS (2006) Stretch increases alveolar epithelial permeability to uncharged micromolecules. Am J Physiol Cell Physiol 290(4):1179. https://doi.org/10.1152/ajpcell.00355.2004

Chancellor MB, Rivas DA, Huang B, Kelly G, Salzman SK (1994) Micturition patterns after spinal trauma as a measure of autonomic functional recovery. J Urol 151(1):250–254. https://doi.org/10.1016/S0022-5347(17)34926-1

Chaqour B, Han JS, Tamura I, Macarak E (2002) Mechanical regulation of IGF-I and IGF-binding protein gene transcription in bladder smooth muscle cells. J Cell Biochem 84(2):264–277. https://doi.org/10.1002/jcb.10013

Chen H, Richard M, Sandler DP, Umbach DM, Kamel F (2007) Head injury and amyotrophic lateral sclerosis. Am J Epidemiol 166(7):810–816

Chen J, Drzewiecki BA, Merryman WD, Pope JC (2013b) Murine bladder wall biomechanics following partial bladder obstruction. J Biomech 46(15):2752–2755. https://www.clinicalkey.es/playcontent/1-s2.0-S0021929013003473. https://doi.org/10.1016/j.jbiomech.2013.07.022

Cheung M, Bao W, Behm DJ et al (2017) Discovery of GSK2193874: an orally active, potent, and selective blocker of transient receptor potential vanilloid 4. ACS Med Chem Lett 8(5):549–554. https://doi.org/10.1021/acsmedchemlett.7b00094

Choo MS, Moon J, Lee HK et al (2017) Mp94-14 urodynamic evaluation of da-8010 in conscious rats with partial bladder outlet obstruction. J Urol 197(4):e1251. https://doi.org/10.1016/j.juro.2017.02.2917

Chopra B, Gever J, Barrick SR et al (2008) Expression and function of rat urothelial P2Y receptors. Am J Physiol Renal Physiol 294(4):F821–F829. http://ajprenal.physiology.org/cgi/content/abstract/294/4/F821. https://doi.org/10.1152/ajprenal.00321.2006

Chu EK, Foley JS, Cheng J, Patel AS, Drazen JM, Tschumperlin DJ (2005) Bronchial epithelial compression regulates epidermal growth factor receptor family ligand expression in an autocrine manner. Am J Respir Cell Mol Biol 32(5):373–380. https://doi.org/10.1165/rcmb.2004-0266OC

Cohen SM, Arnold LL, Uzvolgyi E et al (2002) Possible role of dimethylarsinous acid in dimethylarsinic acid-induced urothelial toxicity and regeneration in the rat. Chem Res Toxicol 15(9):1150–1157. https://doi.org/10.1021/tx020026z

Cohen TS, Gray Lawrence G, Khasgiwala A, Margulies SS (2010) MAPK activation modulates permeability of isolated rat alveolar epithelial cell monolayers following cyclic stretch. PLoS One 5(4):e10385. https://doi.org/10.1371/journal.pone.0010385

Cross MJ, Stewart A, Hodgkin MN, Kerr DJ, Wakelam MJ (1995) Wortmannin and its structural analogue demethoxyviridin inhibit stimulated phospholipase A2 activity in swiss 3T3 cells. wortmannin is not a specific inhibitor of phosphatidylinositol 3-kinase. J Biol Chem 270(43):25352–25355. https://www.ncbi.nlm.nih.gov/pubmed/7592698. https://doi.org/10.1074/jbc.270.43.25352

Dai Y, Tian Y, Luo DY et al (2015) Cyclic stretch induces human bladder smooth muscle cell proliferation in vitro through muscarinic receptors. Mol Med Rep 11(3):2292–2298. https://doi.org/10.3892/mmr.2014.2984

Danciu TE, Adam RM, Naruse K, Freeman MR, Hauschka PV (2003) Calcium regulates the PI3K-akt pathway in stretched osteoblasts. FEBS Lett 536(1):193–197. https://doi.org/10.1016/S0014-5793(03)00055-3

Davidovich N, Huang J, Margulies SS (2013) Reproducible uniform equibiaxial stretch of precision-cut lung slices. Am J Physiol Lung Cell Mol Physiol 304(4):210. https://doi.org/10.1152/ajplung.00224.2012

De Groat WC (1995) Mechanisms underlying the recovery of lower urinary tract function following spinal cord injury. Paraplegia 33(9):493–505

de Groat WC, Yoshimura N (2009) Afferent nerve regulation of bladder function in health and disease. Handb Exp Pharmacol 194:91–138. https://doi.org/10.1007/978-3-540-79090-7_4

Deng L, Fairbank NJ, Fabry B, Smith PG, Maksym GN (2004) Localized mechanical stress induces time-dependent actin cytoskeletal remodeling and stiffening in cultured airway smooth muscle cells. Am J Physiol Cell Physiol 287(2):440. https://doi.org/10.1152/ajpcell.00374.2003

Di Virgilio F (2013) The therapeutic potential of modifying inflammasomes and NOD-like receptors. Pharmacol Rev 65(3):872–905. https://www.ncbi.nlm.nih.gov/pubmed/23592611. https://doi.org/10.1124/pr.112.006171

DiLeonardi AM, Huh JW, Raghupathi R (2009) Impaired axonal transport and neurofilament compaction occur in separate populations of injured axons following diffuse brain injury in the immature rat. Brain Res 1263:174–182

Du S, Araki I, Mikami Y et al (2007) Amiloride-sensitive ion channels in urinary bladder epithelium involved in mechanosensory transduction by modulating stretch-evoked adenosine triphosphate release. Urology (Ridgewood, N.J.) 69(3):590–595. https://www.clinicalkey.es/playcontent/1-s2.0-S0090429507001021. https://doi.org/10.1016/j.urology.2007.01.039

Dunton CL, Purves JT, Hughes FM, Jin H, Nagatomi J (2018) Elevated hydrostatic pressure stimulates ATP release which mediates activation of the NLRP3 inflammasome via P2X4 in rat urothelial cells. Int Urol Nephrol 50(9):1607–1617. https://doi.org/10.1007/s11255-018-1948-0

Engler AJ, Sen S, Sweeney HL, Discher DE (2006) Matrix elasticity directs stem cell lineage specification. Cell 126(4):677–689

Ewalt DH, Howard PS, Blyth B et al (1992) Is lamina propria matrix responsible for normal bladder compliance? J Urol 148(2):544–549. https://doi.org/10.1016/S0022-5347(17)36650-8

Farrugia G, Holm AN, Rich A, Sarr MG, Szurszewski JH, Rae JL (1999) A mechanosensitive calcium channel in human intestinal smooth muscle cells. Gastroenterology (New York, N.Y. 1943) 117(4):900–905. https://doi.org/10.1016/S0016-5085(99)70349-5

Ferguson DR, Kennedy I, Burton TJ (1997) ATP is released from rabbit urinary bladder epithelial cells by hydrostatic pressure changes—a possible sensory mechanism? J Physiol 505(Pt 2):503–511. http://jp.physoc.org/content/505/Pt_2/503.abstract. https://doi.org/10.1111/j.1469-7793.1997.503bb.x

Fewell JE, Hislop AA, Kitterman JA, Johnson P (1983) Effect of tracheostomy on lung development in fetal lambs. J Appl Physiol Respir Environ Exerc Physiol 55(4):1103–1108. https://doi.org/10.1152/jappl.1983.55.4.1103

Fischer RS, Myers KA, Gardel ML, Waterman CM (2012) Stiffness-controlled three-dimensional extracellular matrices for high-resolution imaging of cell behavior. Nat Protoc 7(11):2056–2066

Fisher JL, Margulies SS (2002) Na(+)-K(+)-ATPase activity in alveolar epithelial cells increases with cyclic stretch. Am J Physiol Lung Cell Mol Physiol 283(4):737. https://doi.org/10.1152/ajplung.00030.2001

Fleminger S, Oliver DL, Lovestone S, Rabe-Hesketh S, Giora A (2003) Head injury as a risk factor for alzheimer's disease: the evidence 10 years on; a partial replication. J Neurol Neurosurg Psychiatry 74(7):857–862

Franceschini A, Capece M, Chiozzi P et al (2015) The P2X7 receptor directly interacts with the NLRP3 inflammasome scaffold protein. FASEB J 29(6):2450–2461. https://doi.org/10.1096/fj.14-268714

Friedrich EE, Hong Z, Xiong S et al (2019) Endothelial cell Piezo1 mediates pressure-induced lung vascular hyperpermeability via disruption of adherens junctions. Proc Natl Acad Sci 116(26):12980–12985. https://www.jstor.org/stable/26744131. https://doi.org/10.1073/pnas.1902165116

Fusco F, Creta M, De Nunzio C et al (2018) Progressive bladder remodeling due to bladder outlet obstruction: a systematic review of morphological and molecular evidences in humans. BMC Urol 18(1):15–14. https://doi.org/10.1186/s12894-018-0329-4

Galvin DJ, Watson RWG, Gillespie JI, Brady H, Fitzpatrick JM (2002) Mechanical stretch regulates cell survival in human bladder smooth muscle cells in vitro. Am J Physiol Ren Physiol 283(6):1192–1199. http://ajprenal.physiology.org/content/283/6/F1192. https://doi.org/10.1152/ajprenal.00168.2002

Gao X, Chen J (2011) Mild traumatic brain injury results in extensive neuronal degeneration in the cerebral cortex. J Neuropathol Exp Neurol 70(3):183–191

Geng H, Zhao L, Liang Z et al (2015) ERK5 positively regulates cigarette smoke-induced urocystic epithelial-mesenchymal transition in SV-40 immortalized human urothelial cells. Oncol Rep 34(3):1581–1588. https://www.ncbi.nlm.nih.gov/pubmed/26177962. https://doi.org/10.3892/or.2015.4130

Gevaert T, Vandepitte J, Ost D, Nilius B, De Ridder D (2007) Autonomous contractile activity in the isolated rat bladder is modulated by a TRPV1 dependent mechanism. Neurourol Urodyn 26(3):424–432. https://api.istex.fr/ark:/67375/WNG-0RDPSMF8-1/fulltext.pdf. https://doi.org/10.1002/nau.20313

Grailer JJ, Canning BA, Kalbitz M et al (2014) Critical role for the NLRP3 inflammasome during acute lung injury. J Immunol (1950) 192(12):5974–5983. https://www.ncbi.nlm.nih.gov/pubmed/24795455. https://doi.org/10.4049/jimmunol.1400368

Gurwitz D, Cunningham DD (1988) Thrombin modulates and reverses neuroblastoma neurite outgrowth. Proc Natl Acad Sci U.S 85(10):3440–3444

Haberstroh KM, Kaefer M, Retik AB, Freeman MR, Bizios R (1999) The effects of sustained hydrostatic pressure on select bladder smooth muscle cell functions. J Urol 162(6):2114–2118. https://doi.org/10.1016/S0022-5347(05)68136-0

Haberstroh KM, Kaefer M, Bizios R (2000) Inhibition of pressure induced bladder smooth muscle cell hyperplasia using CRM197. J Urol 164(4):1329–1333. https://doi.org/10.1016/S0022-5347(05)67189-3

Halachmi S, Aitken KJ, Szybowska M et al (2006) Role of signal transducer and activator of transcription 3 (STAT3) in stretch injury to bladder smooth muscle cells. Cell Tissue Res 326(1):149–158. https://www.ncbi.nlm.nih.gov/pubmed/16705451. https://doi.org/10.1007/s00441-006-0204-6

Han JH, Yu HS, Lee JY et al (2015) Simple modification of the bladder outlet obstruction index for better prediction of endoscopically-proven prostatic obstruction: a preliminary study. PLoS One 10(10):e0141745. https://www.ncbi.nlm.nih.gov/pubmed/26505196. https://doi.org/10.1371/journal.pone.0141745

Harding R, Hooper SB (1996) Regulation of lung expansion and lung growth before birth. J Appl Physiol (1985) 81(1):209–224. https://doi.org/10.1152/jappl.1996.81.1.209

Hasaneen NA, Zucker S, Cao J, Chiarelli C, Panettieri RA, Foda HD (2005) Cyclic mechanical strain-induced proliferation and migration of human airway smooth muscle cells: role of EMMPRIN and MMPs. FASEB J 19(11):1507–1509. https://doi.org/10.1096/fj.04-3350fje

Hawryluk GWJ, Citerio G, Hutchinson P et al (2022) Intracranial pressure: current perspectives on physiology and monitoring. Intensive Care Med 48(10):1471–1481

Hirose M, Ishizaki T, Watanabe N et al (1998) Molecular dissection of the rho-associated protein kinase (p160ROCK)-regulated neurite remodeling in neuroblastoma N1E-115 cells. J Cell Biol 141(7):1625–1636

Hughes J, Francis M, Vivar NP, Kennis JG et al (2014) Inflammasomes are important mediators of cyclophosphamide-induced bladder inflammation. Am J Physiol Renal Physiol 306(3):F299–F308. https://www.ncbi.nlm.nih.gov/pubmed/24285499. https://doi.org/10.1152/ajprenal.00297.2013

Hughes FM, Hill HM, Wood CM et al (2016) The NLRP3 inflammasome mediates inflammation produced by bladder outlet obstruction. J Urol 195(5):1598–1605. https://www.clinicalkey.es/playcontent/1-s2.0-S0022534715054439. https://doi.org/10.1016/j.juro.2015.12.068

Hughes J, Francis M, Sexton SJ, Jin H, Govada V, Purves JT (2017) Bladder fibrosis during outlet obstruction is triggered through the NLRP3 inflammasome and the production of IL-1β. Am J Physiol Renal Physiol 313(3):F603–F610. https://www.ncbi.nlm.nih.gov/pubmed/28592436. https://doi.org/10.1152/ajprenal.00128.2017

Hughes J, Francis M, Sexton SJ, Ledig PD, Yun CE, Jin H, Purves JT (2019) Bladder decompensation and reduction in nerve density in a rat model of chronic bladder outlet obstruction are attenuated with the NLRP3 inhibitor glyburide. Am J Physiol Renal Physiol 316(1):F113–F120. https://www.ncbi.nlm.nih.gov/pubmed/30353742. https://doi.org/10.1152/ajprenal.00400.2018

Iguchi N, Hou A, Koul HK, Wilcox DT (2014) Partial bladder outlet obstruction in mice may cause E-cadherin repression through hypoxia induced pathway. J Urol 192(3):964–972. https://www.clinicalkey.es/playcontent/1-s2.0-S002253471402970X. https://doi.org/10.1016/j.juro.2014.03.037

Ihara T, Mitsui T, Nakamura Y et al (2018) The oscillation of intracellular ca(2+) influx associated with the circadian expression of Piezo1 and TRPV4 in the bladder urothelium. Sci Rep 8(1):5699-w. https://doi.org/10.1038/s41598-018-23115-w

Irwin DE, Kopp ZS, Agatep B, Milsom I, Abrams P (2011) Worldwide prevalence estimates of lower urinary tract symptoms, overactive bladder, urinary incontinence and bladder outlet obstruction. BJU Int 108(7):1132–1138. https://doi.org/10.1111/j.1464-410X.2010.09993.x

Islam SS, Mokhtari RB, El Hout Y et al (2014) TGF-β1 induces EMT reprogramming of porcine bladder urothelial cells into collagen producing fibroblasts-like cells in a Smad2/Smad3-dependent manner. J Cell Commun Signal 8(1):39–58. https://doi.org/10.1007/s12079-013-0216-4

Jafari S, Etminan M, Aminzadeh F, Samii A (2013) Head injury and risk of parkinson disease: a systematic review and meta-analysis. Mov Disord 28(9):1222–1229

Janssen DAW, Schalken JA, Heesakkers JPFA (2017) Urothelium update: how the bladder mucosa measures bladder filling. Acta Physiol (Oxf) 220(2):201–217. https://doi.org/10.1111/apha.12824

Jaslove JM, Goodwin K, Sundarakrishnan A et al (2022) Transmural pressure signals through retinoic acid to regulate lung branching. *Development* 149(2):dev199726. https://doi.org/10.1242/dev.199726. Epub 2022 Jan 20

Kadam R, Wiafe B, Metcalfe PD (2021) Mesenchymal stem cells ameliorate partial bladder outlet obstruction-induced epithelial-mesenchymal transition type II independent of mast cell recruitment and degranulation. Can Urol Assoc J 15(1):E29–E35. https://www.ncbi.nlm.nih.gov/pubmed/32701447. https://doi.org/10.5489/cuaj.6501

Karim OMA, Cendron M, Mostwin JL, Gearhart JP (1993) Developmental alterations in the fetal lamb bladder subjected to partial urethral obstruction in utero. J Urol 150(3):1060–1063. https://doi.org/10.1016/S0022-5347(17)35686-0

Kawamata H, Kameyama S, Nan L, Kawai K, Oyasu R (1993) Effect of epidermal growth factor and transforming growth factor β1 on growth and invasive potentials of newly established rat bladder carcinoma cell lines. Int J Cancer 55(6):968–973. https://doi.org/10.1002/ijc.2910550616

Kim JC, Yoon JY, Seo SI, Hwang TK, Park YH (2000) Effects of partial bladder outlet obstruction and its relief on types I and III collagen and detrusor contractility in the rat. Neurourol Urodyn 19(1):29–42. https://api.istex.fr/ark:/67375/WNG-09DHVJ1P-C/fulltext.pdf.

Kim S, Jin J, Kunapuli SP (2004) Akt activation in platelets depends on gi signaling pathways. J Biol Chem 279(6):4186–4195. https://doi.org/10.1074/jbc.M306162200

Korossis S, Bolland F, Ingham E, Fisher J, Kearney J, Southgate J (2006) Review: tissue engineering of the urinary bladder: considering structure-function relationships and the role of mechanotransduction. Tissue Eng 12(4):635–644. https://doi.org/10.1089/ten.2006.12.635

Kropp BP, Zhang Y, Tomasek JJ et al (1999) Characterization of cultured bladder smooth muscle cells: assessment of in vitro contractility. J Urol 162(5):1779–1784. https://doi.org/10.1016/S0022-5347(05)68237-7

Kumar V, Chapple CC, Chess-Williams R (2004) Characteristics of adenosine triphosphate [corrected] release from porcine and human normal bladder. J Urol 172(2):744–747. https://www.ncbi.nlm.nih.gov/pubmed/15247774

Kurzrock EA, Lieu DK, de Graffenried LA, Isseroff RR (2005) Rat urothelium: improved techniques for serial cultivation, expansion, freezing and reconstitution onto acellular matrix. J Urol 173(1):281–285. https://doi.org/10.1097/01.ju.0000141585.17953.fa

Lamkanfi M, Mueller JL, Vitari AC et al (2009) Glyburide inhibits the cryopyrin/Nalp3 inflammasome. J Cell Biol 187(1):61–70. https://agris.fao.org/agris-search/search.do?recordID=US201301688892. https://doi.org/10.1083/jcb.200903124

Lee T, Hedlund P, Newgreen D, Andersson K (2007) Urodynamic effects of a novel EP1 receptor antagonist in normal rats and rats with bladder outlet obstruction. J Urol 177(4):1562–1567. https://www.clinicalkey.es/playcontent/1-s2.0-S0022534706031090. https://doi.org/10.1016/j.juro.2006.11.070

Lee JK, Lu S, Madhukar A (2010) Real-time dynamics of Ca2+, caspase-3/7, and morphological changes in retinal ganglion cell apoptosis under elevated pressure. PLoS One 5(10):e13437

Lewis SA (2000) Everything you wanted to know about the bladder epithelium but were afraid to ask. Am J Physiol—Ren Physiol 278(6):867–874. http://ajprenal.physiology.org/content/278/6/F867. https://doi.org/10.1152/ajprenal.2000.278.6.f867

Li C, Rezania S, Kammerer S et al (2015) Piezo1 forms mechanosensitive ion channels in the human MCF-7 breast cancer cell line. Sci Rep 5(1):8364. https://www.ncbi.nlm.nih.gov/pubmed/25666479. https://doi.org/10.1038/srep08364

Li W, Bao G, Chen W et al (2018) Connexin 43 hemichannel as a novel mediator of sterile and infectious inflammatory diseases. Sci Rep 8(1):166. https://www.ncbi.nlm.nih.gov/pubmed/29317708. https://doi.org/10.1038/s41598-017-18452-1

Lindstroêm S, Mazieres L (1991) Effect of menthol on the bladder cooling reflex in the cat. Acta Physiologica Scandinavica 141(1):1–10. https://api.istex.fr/ark:/67375/WNG-LSSPT8WF-4/fulltext.pdf. https://doi.org/10.1111/j.1748-1716.1991.tb09037.x

Liu M, Skinner SJ, Xu J, Han RN, Tanswell AK, Post M (1992) Stimulation of fetal rat lung cell proliferation in vitro by mechanical stretch. Am J Phys 263(3 Pt 1):376. https://doi.org/10.1152/ajplung.1992.263.3.L376

Liu H, Li W, Chen Z et al (2013) Expression of the NLRP3 inflammasome in cerebral cortex after traumatic brain injury in a rat model. Neurochem Res 38(10):2072–2083. https://doi.org/10.1007/s11064-013-1115-z

Longair MH, Baker DA, Armstrong JD (2011) Simple neurite tracer: open source software for reconstruction, visualization and analysis of neuronal processes. Bioinf (Oxf) 27(17):2453–2454

Maksym GN, Deng L, Fairbank NJ, Lall CA, Connolly SC (2005) Beneficial and harmful effects of oscillatory mechanical strain on airway smooth muscle. Can J Physiol Pharmacol 83(10):913–922. https://doi.org/10.1139/y05-091

Malkowicz SB, Wein AJ, Elbadawi A, Arsdalen KV, Ruggieri MR, Levin RM (1986) Acute biochemical and functional alterations in the partially obstructed rabbit urinary bladder. J Urol 136(6):1324–1329. https://doi.org/10.1016/S0022-5347(17)45331-6

Matsumoto-Miyai K, Kagase A, Murakawa Y, Momota Y, Kawatani M (2009) Extracellular Ca2+ regulates the stimulus-elicited ATP release from urothelium. Auton Neurosci 150(1):94–99. https://www.clinicalkey.es/playcontent/1-s2.0-S1566070209003762. https://doi.org/10.1016/j.autneu.2009.05.253

McGuire EJ, Woodside JR, Borden TA, Weiss RM (2002) Prognostic value of urodynamic testing in myelodysplastic patients. J Urol 167(2):1049–1053. https://doi.org/10.1016/S0022-5347(02)80338-X

McKinley M, O'Loughlin V (2005) Human anatomy, 5th edn. McGrow-Hill

McLatchie LM, Fry CH (2015) ATP release from freshly isolated Guinea-pig bladder urothelial cells: a quantification and study of the mechanisms involved. BJU Int 115(6):987–993. https://doi.org/10.1111/bju.12954

Metcalfe PD, Wang J, Jiao H et al (2010) Bladder outlet obstruction: progression from inflammation to fibrosis. BJU Int 106(11):1686–1694. https://doi.org/10.1111/j.1464-410X.2010.09445.x

Michishita M, Yano K, Tomita K, Matsuzaki O, Kasahara K (1973) Piezo1 expression increases in rat bladder after partial bladder outlet obstruction. Life Sci 2016(166):1–7. https://doi.org/10.1016/j.lfs.2016.10.017

Miyamoto T, Mochizuki T, Nakagomi H et al (2014) Functional role for Piezo1 in stretch-evoked ca^{2+} influx and ATP release in urothelial cell cultures. J Biol Chem 289(23):16565–16575. https://doi.org/10.1074/jbc.M113.528638

Mochizuki T, Sokabe T, Araki I et al (2009) The TRPV4 cation channel mediates stretch-evoked Ca2+ influx and ATP release in primary urothelial cell cultures. J Biol Chem 284(32):21257–21264. https://doi.org/10.1074/jbc.M109.020206

Monnier PP, Sierra A, Schwab JM, Henke-Fahle S, Mueller BK (2003) The rho/ROCK pathway mediates neurite growth-inhibitory activity associated with the chondroitin sulfate proteoglycans of the CNS glial scar. Mol Cell Neurosci 22(3):319–330

Morales-Orcajo E, Siebert T, Böl M (2018) Location-dependent correlation between tissue structure and the mechanical behaviour of the urinary bladder. Acta Biomater 75:263–278. https://doi.org/10.1016/j.actbio.2018.05.014

Mortimer JA, van Duijn CM, Chandra V et al (1991) Head trauma as a risk factor for alzheimer's disease: a collaborative re-analysis of case-control studies. EURODEM risk factors research group. Int J Epidemiol 20(Suppl 2):S28–S35

Naga Prasad SV, Esposito G, Mao L, Koch WJ, Rockman HA (2000) Gbetagamma-dependent phosphoinositide 3-kinase activation in hearts with in vivo pressure overload hypertrophy. J Biol Chem 275(7):4693–4698. https://www.ncbi.nlm.nih.gov/pubmed/10671499

Nagatomi J, Arulanandam BP, Metzger DW, Meunier A, Bizios R (2001) Frequency- and duration-dependent effects of cyclic pressure on select bone cell functions. Tissue Eng 7(6):717–728. https://www.ncbi.nlm.nih.gov/pubmed/11749729. https://doi.org/10.1089/107632701753337672

Nagatomi J, Gloeckner DC, Chancellor MB, DeGroat WC, Sacks MS (2004) Changes in the biaxial viscoelastic response of the urinary bladder following spinal cord injury. Ann Biomed Eng 32(10):1409–1419. https://www.ncbi.nlm.nih.gov/pubmed/15535058. https://doi.org/10.1114/B:ABME.0000042228.89106.48

Nagatomi J, Toosi KK, Chancellor MB, Sacks MS (2008) Contribution of the extracellular matrix to the viscoelastic behavior of the urinary bladder wall. Biomech Model Mechanobiol 7(5):395–404. https://doi.org/10.1007/s10237-007-0095-9

Nagle AS, Klausner AP, Varghese J et al (2017) Quantification of bladder wall biomechanics during urodynamics: a methodologic investigation using ultrasound. J Biomech 61:232–241. https://doi.org/10.1016/j.jbiomech.2017.07.028

Nakagomi H, Yoshiyama M, Mochizuki T et al (2016) Urothelial ATP exocytosis: regulation of bladder compliance in the urine storage phase. Sci Rep 6(1):29761. https://www.ncbi.nlm.nih.gov/pubmed/27412485. https://doi.org/10.1038/srep29761

Nascimento MG, Suzuki S, Wei M et al (2008) Cytotoxicity of combinations of arsenicals on rat urinary bladder urothelial cells in vitro. Toxicology (Amsterdam) 249(1):69–74. https://www.clinicalkey.es/playcontent/1-s2.0-S0300483X08001753. https://doi.org/10.1016/j.tox.2008.04.007

Ndubuizu O, LaManna JC (2007) Brain tissue oxygen concentration measurements. Antioxid Redox Signal 9(8):1207–1220

Negoro H, Urban-Maldonado M, Liou LS, Spray DC, Thi MM, Suadicani SO (2014) Pannexin 1 channels play essential roles in urothelial mechanotransduction and intercellular signaling. PLoS One 9(8):e106269. https://doi.org/10.1371/journal.pone.0106269

Nelson CM, Gleghorn JP, Pang MF et al (2017) Microfluidic chest cavities reveal that transmural pressure controls the rate of lung development. Development 144(23):4328–4335. https://doi.org/10.1242/dev.154823

Nguyen HT, Adam RM, Bride SH, Park JM, Peters CA, Freeman MR (2000) Cyclic stretch activates p38 SAPK2-, ErbB2-, and AT1-dependent signaling in bladder smooth muscle cells. Am J Physiol Cell Physiol 279(4):1155–1167. http://ajpcell.physiology.org/content/279/4/C1155. https://doi.org/10.1152/ajpcell.2000.279.4.c1155

Oka M, Fukui T, Ueda M, Tagaya M, Oyama T, Tanaka M (2009) Suppression of bladder oxidative stress and inflammation by a phytotherapeutic agent in a rat model of partial bladder outlet obstruction. J Urol 182(1):382–390. https://www.clinicalkey.es/playcontent/1-s2.0-S0022534709005254. https://doi.org/10.1016/j.juro.2009.02.104

Olsen SM, Stover JD, Nagatomi J (2011) Examining the role of mechanosensitive ion channels in pressure mechanotransduction in rat bladder urothelial cells. Ann Biomed Eng 39(2):688–697. https://doi.org/10.1007/s10439-010-0203-3

Orlichenko LS, Radisky DC (2008) Matrix metalloproteinases stimulate epithelial-mesenchymal transition during tumor development. Clin Exp Metastasis 25(6):593–600. https://doi.org/10.1007/s10585-008-9143-9

Orsola A, Adam RM, Peters CA, Freeman MR (2002) The decision to undergo DNA or protein synthesis is determined by the degree of mechanical deformation in human bladder muscle cells. Urology (Ridgewood, N.J.) 59(5):779–783. https://doi.org/10.1016/S0090-4295(01)01648-X

Pandita RK, Fujiwara M, Alm P, Andersson K (2000) Cystometric evaluation of bladder function in non-anesthetized mice with and without bladder outlet obstruction. J Urol 164(4):1385–1389. https://doi.org/10.1016/S0022-5347(05)67204-7

Park JM, Borer JG, Freeman MR, Peters CA (1998) Stretch activates heparin-binding EGF-like growth factor expression in bladder smooth muscle cells. Am J Phys 275(5):1247. https://doi.org/10.1152/ajpcell.1998.275.5.C1247

Park JM, Adam RM, Peters CA et al (1999a) AP-1 mediates stretch-induced expression of HB-EGF in bladder smooth muscle cells. Am J Phys 277(2):294. https://doi.org/10.1152/ajpcell.1999.277.2.C294

Park JM, Yang T, Arend LJ et al (1999b) Obstruction stimulates COX-2 expression in bladder smooth muscle cells via increased mechanical stretch. Am J Physiol Ren Physiol 276(1):129–136. http://ajprenal.physiology.org/content/276/1/F129. https://doi.org/10.1152/ajprenal.1999.276.1.F129

Pelegrin P, Surprenant A (2006) Pannexin-1 mediates large pore formation and interleukin-1β release by the ATP-gated P2X7 receptor. EMBO J 25(21):5071–5082. https://api.istex.fr/ark:/67375/WNG-60HKM6NV-J/fulltext.pdf. https://doi.org/10.1038/sj.emboj.7601378

Peters CA, Vasavada S, Dator D et al (1992) The effect of obstruction on the developing bladder. J Urol 148(2):491–496. https://doi.org/10.1016/S0022-5347(17)36638-7

Pétrilli V, Papin S, Dostert C, Mayor A, Martinon F, Tschopp J (2007) Activation of the NALP3 inflammasome is triggered by low intracellular potassium concentration. Cell Death Differ 14(9):1583–1589. https://doi.org/10.1038/sj.cdd.4402195

Prado GR, Ross JD, DeWeerth SP, LaPlaca MC (2005) Mechanical trauma induces immediate changes in neuronal network activity. J Neural Eng 2(4):148–158

Qu M, Liu B, Wang H, Yan Z, Shen B, Jiang Z (2007) Frequency-dependent phenotype modulation of vascular smooth muscle cells under cyclic mechanical strain. J Vasc Res 44(5):345–353. https://www.karger.com/Article/Abstract/102278. https://doi.org/10.1159/000102278

Quan R, Liang W, Li H, Ning Q, Shang D (2022) Silencing of miR-10b-5p alleviates the mechanical stretch-induced proliferation of HASMCs. Tissue Cell 74:101700. https://doi.org/10.1016/j.tice.2021.101700

Radio NM, Breier JM, Shafer TJ, Mundy WR (2008) Assessment of chemical effects on neurite outgrowth in PC12 cells using high content screening. Toxicol Sci 105(1):106–118

Rangel-Castillo L, Gopinath S, Robertson CS (2008) Management of intracranial hypertension. *Neurologic Clinics* 26(2):521–541

Rashid B, Destrade M, Gilchrist MD (2013) Mechanical characterization of brain tissue in simple shear at dynamic strain rates. J Mech Behav Biomed Mater 28:71–85

Reigada D, Lu W, Zhang M, Mitchell CH (2008) Elevated pressure triggers a physiological release of ATP from the retina: possible role for pannexin hemichannels. Neuroscience 157(2):396–404. https://www.clinicalkey.es/playcontent/1-s2.0-S0306452208012487. https://doi.org/10.1016/j.neuroscience.2008.08.036

Rensen SSM, Doevendans PA, van Eys GJJM (2007) Regulation and characteristics of vascular smooth muscle cell phenotypic diversity. Neth Hear J 15(3):100–108. https://www.narcis.nl/publication/RecordID/oai:cris.maastrichtuniversity.nl:publications%2Fc868ab3c-bb00-442e-9baf-6f03cd9de4e2. https://doi.org/10.1007/BF03085963

Richter K, Kiefer KP, Grzesik BA, Clauss WG, Fronius M (2014) Hydrostatic pressure activates ATP-sensitive K+ channels in lung epithelium by ATP release through pannexin and connexin hemichannels. FASEB J 28(1):45–55. https://doi.org/10.1096/fj.13-229252

Riteau N, Gasse P, Ryffel B et al (2010) Extracellular ATP is a danger signal activating P2X7 receptor in lung inflammation and fibrosis. Am J Respir Crit Care Med 182(6):774–783. https://www.ncbi.nlm.nih.gov/pubmed/20522787. https://doi.org/10.1164/rccm.201003-0359OC

Roby T, Olsen S, Nagatomi J (2008) Effect of sustained tension on bladder smooth muscle cells in three-dimensional culture. Ann Biomed Eng 36(10):1744–1751. https://doi.org/10.1007/s10439-008-9545-5

Roose-Girma M, Lee WP, Mariathasan S et al (2006) Cryopyrin activates the inflammasome in response to toxins and ATP. Nature 440(7081):228–232. https://doi.org/10.1038/nature04515

Rosenblatt N, Hu S, Suki B, Wang N, Stamenović D (2007) Contributions of the active and passive components of the cytoskeletal prestress to stiffening of airway smooth muscle cells. Ann Biomed Eng 35(2):224–234. https://doi.org/10.1007/s10439-006-9228-z

Sano T, Kobayashi T, Negoro H et al (2016) Intravital imaging of mouse urothelium reveals activation of extracellular signal-regulated kinase by stretch-induced intravesical release of ATP. Physiol Rep 4(21):np-n/a. https://doi.org/10.14814/phy2.13033

Sayas CL, Ariaens A, Ponsioen B, Moolenaar WH (2006) GSK-3 is activated by the tyrosine kinase Pyk2 during LPA1-mediated neurite retraction. Mol Biol Cell 17(4):1834–1844

Schittny JC, Miserocchi G, Sparrow MP (2000) Spontaneous peristaltic airway contractions propel lung liquid through the bronchial tree of intact and fetal lung explants. Am J Respir Cell Mol Biol 23(1):11–18. https://doi.org/10.1165/ajrcmb.23.1.3926

Schroder K, Tschopp J (2010) The inflammasomes. Cell 140(6):821–832. https://doi.org/10.1016/j.cell.2010.01.040

Schröder A, Kirwan TP, Jiang J, Aitken KJ, Bägli DJ (2013) Rapamycin attenuates bladder hypertrophy during long-term outlet obstruction in vivo: Tissue, matrix and mechanistic insights. J Urol 189(6):2377–2384. https://www.clinicalkey.es/playcontent/1-s2.0-S0022534713000128. https://doi.org/10.1016/j.juro.2012.12.110

Schubert D, Humphreys S, de Vitry F, Jacob F (1971) Induced differentiation of a neuroblastoma. Dev Biol 25(4):514–546

Schubert D, Heinemann S, Carlisle W et al (1974) Clonal cell lines from the rat central nervous system. Nature (London) 249(5454):224–227

Seeds NW, Gilman AG, Amano T, Nirenberg MW (1970) Regulation of axon formation by clonal lines of a neural tumor. Proc Natl Acad Sci 66(1):160–167

Seki S, Sasaki K, Fraser MO et al (2002) Immunoneutralization of nerve growth factor in lumbosacral spinal cord reduces bladder hyperreflexia in spinal cord injured rats. J Urol 168(5):2269–2274. https://doi.org/10.1016/S0022-5347(05)64369-8

Sellers D, Chess-Williams R, Michel MC (2018) Modulation of lower urinary tract smooth muscle contraction and relaxation by the urothelium. Naunyn Schmiedeberg's Arch Pharmacol 391(7):675–694. https://doi.org/10.1007/s00210-018-1510-8

Sengiku A, Ueda M, Kono J et al (2018) Circadian coordination of ATP release in the urothelium via connexin43 hemichannels. Sci Rep 8(1):1996–1911. https://www.ncbi.nlm.nih.gov/pubmed/29386573. https://doi.org/10.1038/s41598-018-20379-0

Servin-Vences MR, Moroni M, Lewin GR, Poole K (2017) Direct measurement of TRPV4 and PIEZO1 activity reveals multiple mechanotransduction pathways in chondrocytes. elife 2017:6. https://www.ncbi.nlm.nih.gov/pubmed/28135189. https://doi.org/10.7554/eLife.21074

Shabir S, Cross W, Kirkwood LA et al (2013) Functional expression of purinergic P2 receptors and transient receptor potential channels by the human urothelium. Am J Physiol Renal Physiol 305(3):F396–F406. https://www.ncbi.nlm.nih.gov/pubmed/23720349. https://doi.org/10.1152/ajprenal.00127.2013

Sharif-Afshar A, Donohoe JM, Pope JC et al (2005) Stromal hyperplasia in male bladders upon loss of transforming growth factor-β signaling in fibroblasts. discussion. J Urol 174(4):1704–1707

Sheinerman KS, Umansky SR (2013) Early detection of neurodegenerative diseases. Cell Cycle (Georgetown, Tex.) 12(1):1–2

Skinner SJ, Somervell CE, Olson DM (1992) The effects of mechanical stretching on fetal rat lung cell prostacyclin production. Prostaglandins 43(5):413–433. https://doi.org/10.1016/0090-6980(92)90125-d

Smith PG, Janiga KE, Bruce MC (1994) Strain increases airway smooth muscle cell proliferation. Am J Respir Cell Mol Biol 10(1):85–90. https://doi.org/10.1165/ajrcmb.10.1.8292386

Smith PG, Garcia R, Kogerman L (1997) Strain reorganizes focal adhesions and cytoskeleton in cultured airway smooth muscle cells. Exp Cell Res 232(1):127–136. https://doi.org/10.1006/excr.1997.3492

Smith PG, Roy C, Zhang YN, Chauduri S (2003) Mechanical stress increases RhoA activation in airway smooth muscle cells. Am J Respir Cell Mol Biol 28(4):436–442. https://doi.org/10.1165/rcmb.4754

Sobue K, Hayashi K, Nishida W (1998) Molecular mechanism of phenotypic modulation of smooth muscle cells. Horm Res 50(2):15–24. https://www.karger.com/Article/Abstract/53119. https://doi.org/10.1159/000053119

Solis AG, Bielecki P, Steach HR et al (2019) Mechanosensation of cyclical force by PIEZO1 is essential for innate immunity. Nature (London) 573(7772):69–74. https://www.ncbi.nlm.nih.gov/pubmed/31435009. https://doi.org/10.1038/s41586-019-1485-8

Steers WD, Broder SR, Persson K et al (1998) Mechanical stretch increases secretion of parathyroid hormone-related protein by cultured bladder smooth muscle cells. J Urol 160(3 Pt 1):908–912. https://doi.org/10.1016/S0022-5347(01)62831-3

Stegemann JP, Nerem RM (2003) Altered response of vascular smooth muscle cells to exogenous biochemical stimulation in two-and three-dimensional culture. Exp Cell Res 283(2):146–155. https://doi.org/10.1016/S0014-4827(02)00041-1

Stover J, Nagatomi J (2007) Cyclic pressure stimulates DNA synthesis through the PI3K/akt signaling pathway in rat bladder smooth muscle cells. Ann Biomed Eng 35(9):1585–1594. https://www.ncbi.nlm.nih.gov/pubmed/17522977. https://doi.org/10.1007/s10439-007-9331-9

Stoyanov B, Volinia S, Hanck T et al (1995) Cloning and characterization of a G protein-activated human phosphoinositide-3 kinase. Science (American Association for the Advancement of Science) 269(522):690–693. https://search.proquest.com/docview/16841124

Sun Y, Keay S, De Deyne PG, Chai TC (2001) Augmented stretch activated adenosine triphosphate release from bladder uroepithelial cells in patients with interstitial cystitis. J Urol 166(5):1951–1956. https://doi.org/10.1016/S0022-5347(05)65726-6

Sun Y, MaLossi J, Jacobs SC, Chai TC (2002) Effect of doxazosin on stretch-activated adenosine triphosphate release in bladder urothelial cells from patients with benign prostatic hyperplasia. Urology 60(2):351–356. https://doi.org/10.1016/s0090-4295(02)01710-7

Sun Y, Kim N, Yang H, Kim S, Huh S (2011) Lysophosphatidic acid induces neurite retraction in differentiated neuroblastoma cells via GSK-3[beta] activation. Mol Cells 31(5):483

Suzuki M, Mizuno A, Kodaira K, Imai M (2003) Impaired pressure sensation in mice lacking TRPV4. J Biol Chem 278(25):22664–22668. http://www.jbc.org/content/278/25/22664.abstract. https://doi.org/10.1074/jbc.M302561200

Swartz MA, Tschumperlin DJ, Kamm RD, Drazen JM (2001) Mechanical stress is communicated between different cell types to elicit matrix remodeling. Proc Natl Acad Sci USA 98(11):6180–6185. https://doi.org/10.1073/pnas.111133298

Syed S, Karadaghy A, Zustiak S (2015) Simple polyacrylamide-based multiwell stiffness assay for the study of stiffness-dependent cell responses. J Vis Exp 97:52643

Takahashi M (2014) NLRP3 inflammasome as a novel player in myocardial infarction. Int Heart J 55(2):101–105. https://www.jstage.jst.go.jp/article/ihj/55/2/55_13-388/_article/-char/en. https://doi.org/10.1536/ihj.13-388

Tamura I, Rosenbloom J, Macarak E, Chaqour B (2001) Regulation of Cyr61 gene expression by mechanical stretch through multiple signaling pathways. Am J Physiol Cell Physiol 281(5):1524–1532. http://ajpcell.physiology.org/content/281/5/C1524. https://doi.org/10.1152/ajpcell.2001.281.5.C1524

Tang-Schomer MD, Patel AR, Baas PW, Smith DH (2010) Mechanical breaking of microtubules in axons during dynamic stretch injury underlies delayed elasticity, microtubule disassembly, and axon degeneration. FASEB J 24(5):1401–1410

Tekgul S, Yoshino K, Bagli D, Carr MC, Mitchell ME, Yao LY (1996) Collagen types I and III localization by in situ hybridization and immunohistochemistry in the partially obstructed young rabbit bladder. J Urol 156(2):582–586. https://doi.org/10.1016/S0022-5347(01)65757-4

Tök L, Naz Ro Lu M, AC UU, Tök Ö (2014) Elevated hydrostatic pressures induce apoptosis and oxidative stress through mitochondrial membrane depolarization in PC12 neuronal cells: a cell culture model of glaucoma. J Recep Signal Transduct 34(5):410–416

Toyofuku T, Yabuki M, Otsu K, Kuzuya T, Hori M, Tada M (1998) Intercellular calcium signaling via gap junction in connexin-43-transfected cells. J Biol Chem 273(3):1519–1528. https://doi.org/10.1074/jbc.273.3.1519

Tschumperlin DJ, Margulies SS (1998) Equibiaxial deformation-induced injury of alveolar epithelial cells in vitro. Am J Phys 275(6):1173. https://doi.org/10.1152/ajplung.1998.275.6.L1173

Tschumperlin DJ, Oswari J, Margulies AS (2000) Deformation-induced injury of alveolar epithelial cells. effect of frequency, duration, and amplitude. Am J Respir Crit Care Med 162(2 Pt 1):357–362. https://doi.org/10.1164/ajrccm.162.2.9807003

Tschumperlin DJ, Shively JD, Swartz MA et al (2002) Bronchial epithelial compression regulates MAP kinase signaling and HB-EGF-like growth factor expression. Am J Physiol Lung Cell Mol Physiol 282(5):904. https://doi.org/10.1152/ajplung.00270.2001

Tschumperlin DJ, Shively JD, Kikuchi T, Drazen JM (2003) Mechanical stress triggers selective release of fibrotic mediators from bronchial epithelium. Am J Respir Cell Mol Biol 28(2):142–149. https://doi.org/10.1165/rcmb.2002-0121OC

Tschumperlin DJ, Dai G, Maly IV et al (2004) Mechanotransduction through growth-factor shedding into the extracellular space. Nature 429(6987):83–86. https://doi.org/10.1038/nature02543

Vadiakas GP, Banes AJ (1992) Verapamil decreases cyclic load-induced calcium incorporation in ROS 17/2.8 osteosarcoma cell cultures. Matrix (Stuttgart) 12(6):439–447. https://doi.org/10.1016/S0934-8832(11)80088-0

Vlaskovska M, Kasakov L, Rong W et al (2001) P2X3 knock-out mice reveal a major sensory role for urothelially released ATP. J Neurosci 21(15):5670–5677. http://www.jneurosci.org/cgi/content/abstract/21/15/5670. https://doi.org/10.1523/jneurosci.21-15-05670.2001

Wang ECY, Lee J, Johnson JP, Kleyman TR, Bridges R, Apodaca G (2003) Hydrostatic pressure-regulated ion transport in bladder uroepithelium. Am J Physiol–Ren Physiol 285(4):651–663. http://ajprenal.physiology.org/content/285/4/F651. https://doi.org/10.1152/ajprenal.00403.2002

Watanabe T, Rivas DA, Chancellor MB (1996) Urodynamics of spinal cord injury. Urol Clin N Am 23(3):459–474. https://doi.org/10.1016/S0094-0143(05)70325-6

Werner C, Engelhard K (2007) Pathophysiology of traumatic brain injury. Br J Anaesth 99(1):4–9

Westerink RHS, Ewing AG (2008) PC12 cell as model for neurosecretion. Acta Physiol 192(2):273–285

Wirtz HR, Dobbs LG (1990) Calcium mobilization and exocytosis after one mechanical stretch of lung epithelial cells. Science 250(4985):1266–1269. https://doi.org/10.1126/science.2173861
Worth NF, Rolfe BE, Song J, Campbell GR (2001) Vascular smooth muscle cell phenotypic modulation in culture is associated with reorganisation of contractile and cytoskeletal proteins. Cell Motil Cytoskeleton 49(3):130–145. https://api.istex.fr/ark:/67375/WNG-4WP9R0X9-S/fulltext.pdf. https://doi.org/10.1002/cm.1027
Wu H, Mahmood A, Qu C, Xiong Y, Chopp M (2012) Simvastatin attenuates axonal injury after experimental traumatic brain injury and promotes neurite outgrowth of primary cortical neurons. Brain Res 1486:121–130
Xiong Y, Mahmood A, Chopp M (2009) Emerging treatments for traumatic brain injury. Expert Opin Emerg Drugs 14(1):67–84
Yamada T, Ugawa S, Ueda T, Ishida Y, Kajita K, Shimada S (2009) Differential localizations of the transient receptor potential channels TRPV4 and TRPV1 in the mouse urinary bladder. J Histochem Cytochem 57(3):277–287. http://www.jhc.org/cgi/content/abstract/57/3/277. https://doi.org/10.1369/jhc.2008.951962
Yamaguchi O (2004) Response of bladder smooth muscle cells to obstruction: signal transduction and the role of mechanosensors. Urology 63(3 Suppl 1):11–16. https://doi.org/10.1016/j.urology.2003.12.002
Yeung T, Georges PC, Flanagan LA et al (2005) Effects of substrate stiffness on cell morphology, cytoskeletal structure, and adhesion. Cell Motil Cytoskeleton 60(1):24–34
Yu W (2015) Polarized ATP distribution in urothelial mucosal and serosal space is differentially regulated by stretch and ectonucleotidases. Am J Physiol Renal Physiol 309(10):F864–F872. https://www.ncbi.nlm.nih.gov/pubmed/26336160. https://doi.org/10.1152/ajprenal.00175.2015
Yu G, Bo S, Xiyu J, Enqing X (2003) Effect of bladder outlet obstruction on detrusor smooth muscle cell: an in vitro study. J Surg Res 114(2):202–209. https://doi.org/10.1016/s0022-4804(03)00333-0
Yu W, Khandelwal P, Apodaca G (2009) Distinct apical and basolateral membrane requirements for stretch-induced membrane traffic at the apical surface of bladder umbrella cells. Mol Biol Cell 20(1):282–295. http://www.molbiolcell.org/cgi/content/abstract/20/1/282. https://doi.org/10.1091/mbc.E08-04-0439

Postface

As described in the first chapter, hydrostatic pressure is everywhere around us and inside our bodies, wherever there are fluids like gas and liquids, but we do not feel it constantly. When there is a temporal change or a difference between two locations in hydrostatic pressure, we may realize its existence and detect some differences directly or indirectly. We may experience a squeeze on the eardrums while diving underwater or shortness of breath because of low oxygen tension at high altitudes. Low barometric pressures not only cause bad weather but also affect people's health through joint pains and headaches. The cells of our organs are constantly subjected to "changes" in hydrostatic pressure through external factors like altitudes or climate and internal factors like contraction of muscular tissues, filling of hollow organs, or weight bearing. The in-vitro cell culture model studies reviewed or presented in the volume aimed to recapture the pathophysiological changes that occur under altered levels of hydrostatic pressure inside our bodies. All have incorporated a control group of cells not exposed to the elevated hydrostatic pressure stimuli as a reference point to determine the effect of "changes" in hydrostatic pressure, not the absolute hydrostatic pressure that the altitudes and weather may influence. For example, the studies that examined the effect of cyclic pressure on bone cells intended to model the hydrostatic pressure oscillations due to cyclic loading on the skeleton of walking. The key findings were that compared to the control group of cells maintained under atmospheric pressure, the cells exposed to cyclic pressure exhibited behaviors necessary for maintaining healthy bone mass. Although a debate still exists about whether the mechanical adaptation of bone homeostasis is regulated by cyclic hydrostatic pressure loading or oscillatory flow of interstitial fluids, the studies presented in this volume and by other investigators certainly provide pieces of evidence that bone cells like osteoblasts, osteoclasts, and bone marrow cells respond to changes in hydrostatic pressure in ways that impact the physiology of the bone. This idea that cells are responsive to changes in hydrostatic pressure of physiological magnitudes has been proven for many other organ systems, including the vascular, neuronal, pulmonary, and urological organs. One of

J. Nagatomi, *Mechanobiology of Hydrostatic Pressure*, Frontiers of Biomechanics 5, https://doi.org/10.1007/978-981-95-4405-9

the most critical yet unanswered questions is how these cells sense the small changes in hydrostatic pressure.

Some studies presented in this volume examined the hypothesis that membrane-bound stretch-activated ion channels, such as TRP and piezo proteins, mediate hydrostatic pressure stimuli and calcium-dependent cellular responses. The rationale behind this hypothesis was that hydrostatic pressure could directly activate these what was known as stretch-activated channels or indirectly through stretching of the cell membrane due to cell deformation. Because hydrostatic pressure is isotropic and the pressures on the inside and outside of cells are assumed to equilibrate instantaneously, a general understanding is that there is no movement of fluids or solutes across the cell membrane. In addition, cells are considered incompressible as they are traditionally modeled as bags of water; thus, small changes in hydrostatic pressure are not believed to cause any cell deformation. Following these assumptions would lead to a conclusion that hydrostatic pressure might directly activate the membrane-bound stretch-activated ion channels but not through the deformation of cells. However, recent studies and emerging theories suggest cells can change shape and size when subjected to elevated hydrostatic pressure. More specifically, when exposed to hydrostatic pressure in the order of 50 to 70 mmHg, both vascular endothelial cells adhered to a rigid substrate (Yoshino et al. 2020) and leukemia cells in suspension (Hui et al. 2014) decrease volume over 10- to 100-min periods. A separate study showed that lung cancer cells increase cell volume and motility when exposed to 20-mmHg hydrostatic pressure for up to 8 h (Kao et al. 2017). Although there is a difference in responses between the cell types, all the findings are common in that the cell volume changes occurred minutes or hours after hydrostatic pressure exposure. This time scale for the observed cell volume change suggests that deformation under hydrostatic pressure is not mechanical deformation commonly observed in a deformable body under compressive, tensile, or shear loading. Instead, it is a cell volume change due to efflux or influx of water across the cell membrane that results from activation of water channel, Aquaporin (AQP) (Yoshino et al. 2020; Kao et al. 2017), or ionic concentration imbalance between inside and outside the cell (Hui et al. 2014). To date, it is still unknown how hydrostatic pressure changes lead to activation of AQP for water transport or other ion channels for osmolyte transport. Presumably, water and ion fluxes can lead to local changes in hydrostatic pressure that can result in further transport of water and ions to trigger functional changes of the cell. Emerging theories also look to the changes in cytoskeletal tension due to changes in hydrostatic pressure (Hui et al. 2014; Liu et al. 2019; Chugh et al. 2022). Further studies using more advanced technology are needed to measure these small changes in pressure, cytoskeletal tension, and water transport inside and near the cell membrane, in order to make progress in our understanding of hydrostatic pressure mechanobiology.

References

Chugh M, Munjal A, Megason SG (2022) Hydrostatic pressure as a driver of cell and tissue morphogenesis. Semin Cell Dev Biol 131:134–145. https://doi.org/10.1016/j.semcdb.2022.04.021

Hui TH, Zhou ZL, Qian J, Lin Y, Ngan AHW, Gao H (2014) Volumetric deformation of live cells induced by pressure-activated cross-membrane ion transport. Phys Rev Lett 113(11):118101. https://www.ncbi.nlm.nih.gov/pubmed/25260007. https://doi.org/10.1103/PhysRevLett.113.118101

Kao Y, Jheng J, Pan H, Liao W, Lee C, Kuo P (2017) Elevated hydrostatic pressure enhances the motility and enlarges the size of the lung cancer cells through aquaporin upregulation mediated by caveolin-1 and ERK1/2 signaling. Oncogene 36(6):863–874. https://www.ncbi.nlm.nih.gov/pubmed/27499095. https://doi.org/10.1038/onc.2016.255

Liu S, Tao R, Wang M et al (2019) Regulation of cell behavior by hydrostatic pressure. Appl Mech Rev 71(4):408031. https://doi.org/10.1115/1.4043947

Yoshino D, Funamoto K, Sato K, Kenry SM, Lim CT (2020) Hydrostatic pressure promotes endothelial tube formation through aquaporin 1 and ras-ERK signaling. Commun Biol 3(1):152. https://www.ncbi.nlm.nih.gov/pubmed/32242084. https://doi.org/10.1038/s42003-020-0881-9

J. Nagatomi, *Mechanobiology of Hydrostatic Pressure*, Frontiers of Biomechanics 5, https://doi.org/10.1007/978-981-95-4405-9

MIX
Papier aus verantwortungsvollen Quellen
Paper from responsible sources
FSC® C105338

If you have any concerns about our products,
you can contact us on
ProductSafety@springernature.com

In case Publisher is established outside the EU,
the EU authorized representative is:
Springer Nature Customer Service Center GmbH
Europaplatz 3, 69115 Heidelberg, Germany

Printed by Libri Plureos GmbH
in Hamburg, Germany